ANCIENT MINERALOGY;

OR,

AN INQUIRY RESPECTING

MINERAL SUBSTANCES

MENTIONED BY

THE ANCIENTS:

WITH OCCASIONAL REMARKS

ON

THE USES TO WHICH THEY WERE APPLIED.

By N. F. MOORE, LL.D.

Second Edition.

NEW YORK:
HARPER & BROTHERS, PUBLISHERS,
FRANKLIN SQUARE.
1859.

PREFACE TO THE FIRST EDITION.

THE author is aware that in venturing upon a scientific theme he transcends the limits of his province, but he relies for indulgence on the fact that his subject is considered chiefly in the light in which it presents itself in the writings of those ancients whom it is his proper business to interpret.* It is, moreover, possible for one who makes no pretension to original discovery—who, far from being a proficient, is himself a learner, to collect from various quarters, and arrange, what may contribute to the entertainment or instruction of others who want leisure or opportunity for similar pursuits. And if in the performance of this task he occasionally fail, he may reasonably hope to find in those whom superior knowledge qualifies to aid

* The author was professor of the Greek and Latin languages in Columbia College at that time.

him, not severe censors, but indulgent friends. As such, then, does the author look to the learned and scientific, for benevolent suggestions, that may enable him, perhaps, at some future time, to render this, now very imperfect essay, somewhat less unworthy their regard.

PREFACE TO THE SECOND EDITION.

After an interval of four-and-twenty years is brought out this second edition of "Ancient Mineralogy," not because the author thinks there is any call for it, but chiefly with a view to his own personal gratification; that he may enjoy the satisfaction it will afford him to leave his little work in a somewhat more comely and creditable shape. He is aware that the mass of readers will care little about the shape in which a work of this sort, wherein they can feel so little interest, may appear; but, though conscious of its many imperfections, and how little he has accomplished of the much that might be done in the direction he has followed, he will not despair of his book's finding, in a certain class of readers, a considerable number to whom it will be acceptable: to them he would recommend it, humbly hoping that they will read it with the indulgence it requires.

ANCIENT MINERALOGY.

ONE who professes his intention to treat of Ancient Mineralogy may well anticipate some objection to the term, and that it will perhaps be asked, "What is meant by *Ancient Mineralogy?* Mineralogy is a science with which the *ancients* were wholly unacquainted." And certainly they were so, if we understand—what the term now implies—a well-digested classification and systematic arrangement of all known substances that compose the mineral world. But, in regard as well to Mineralogy as to Botany, Chemistry, and other branches of Natural Science, it will hardly be denied that the ancients possessed much and various information; although, on the other hand, it may be granted that they had little *science*, according to the modern acceptation of the term. Aristotle was a good zoologist, and Theophrastus a good botanist and mineralogist, notwithstanding their ignorance of modern systems; and they fail to obtain, in these characters, the credit they deserve, because the multitude of facts known to them from observation and experience were not systematically connected and completed by the addition of others derived from analogical reasoning, farther experiment, and just induction.

Mineralogy now claims with justice to be regarded as a science, and, in fact, as one of the three grand divisions of Natural History, it holds an elevated rank, all inorganic bodies—the whole mineral kingdom—belonging to its domain.

In accumulating the varied objects of his science, the mineralogist has explored almost every region of the globe, the recesses of every cavern, the depths of every mine, the heart of every rock into which modern enterprise has pierced. The profusion of minerals thus brought to light has been submitted to analysis, and their true composition, as ascertained by the result, has furnished the basis generally adopted for their scientific arrangement. Some writers, however, have been much guided in their classification of minerals by external or physical characters collectively considered; while others, again, have regarded especially that most remarkable and constant of these physical characters, their crystalline form.

In the writings of the ancients we should look in vain for any such well-grounded comprehensive system. They were, for the most part, content with describing individually the minerals they knew, by some more striking external characters, or some obvious property, or by the uses to which they were respectively applied. And in this branch of natural science, as well as in others, there is often great difficulty in ascertaining to what species objects mentioned by the ancients are to be referred, because they were apt to dwell upon features, striking, perhaps, in the individual before them, but in no wise characteristic of its kind. And they were, besides, so poetical often and imaginative that they could hardly bring themselves to give a plain unvarnished description of any thing remarkable, whether in nature or in art.[1]

---

[1] Every page of Pliny's great work, though one in which we might least expect to meet with poetry, will supply examples. Speaking, for instance, of the mysterious virtue of the magnet, he exclaims, "What is more sluggish than the rigid stone? Lo! Nature hath given to it hands and feeling. What more stubborn than the obdurate iron? But it yields and becomes gentle; for, attracted by the magnetic stone, that substance which is the subduer of all things else

But, in justice to the ancients, we should bear in mind that the *science* of mineralogy is, even now, of very recent date; that it is only within the present century that any work has appeared containing more philosophical views of the subject than were exhibited in the little treatise of Theophrastus above two thousand years ago. We know that the ancients were acquainted with the external characters, at least, of a vast number of minerals; with the uses and properties of very many; and we find the only two ancient authors, who treat of them professedly, making various attempts at classification. Why, then, may we not speak of ancient mineralogy as well as of the mineralogy of the last century? If we should light in Pompeii on a cabinet of minerals, which would be one of Pliny's time, it might be found, perhaps, full as well arranged as was that of Sir John Hill, the translator of Theophrastus, sixty years ago.

Besides the descriptions given by Dioscorides[1] of many mineral substances, which form

flies toward the unseen power, and when near stands still, is held, and clings to its embrace."—H. N., xxxvi., 25.

[1] Whom Beckmann (Hist. of Inv., iii., 73) pronounces a good judge of minerals.

an important portion of his materia medica, and the incidental notices of them scattered through the physicians, philosophers, historians, geographers and poets of antiquity, there remains a short treatise of Theophrastus, expressly "Upon Stones," and the last five books of Pliny's great work are chiefly devoted to the consideration of minerals.

The little book of Theophrastus appears to have formed part of some larger work; in which he may, perhaps, have treated more fully of the character and properties of minerals. In the piece which has survived to us his chief endeavor seems to be the settlement of some principle of classification; and, though he introduces a considerable number of mineral substances, it is often rather by way of illustration than with a view to characterize and describe them; and he evidently refers to them in most cases as to substances with which his reader is acquainted.

Beckmann pronounces Theophrastus "the ablest and most accurate mineralogist of the ancients," thus doing justice to the philosophical character of his work, and admitting that the ancients possessed such an acquaintance with minerals as might be styled *mineralogy.*

Throughout the work of Theophrastus we can perceive that he is striving, in a very philosophical spirit, to lay down some satisfactory basis of arrangement; but the appropriate crystalline forms of minerals not having been attended to, except in some few instances, and their chemical composition being totally unknown, his efforts are, of course, not entirely successful. If we view in this light Theophrastus's work, and bear in mind that it was not his intention to describe, nor even to name all the then known minerals, we shall less wonder at his brevity, and at the slight and merely passing notice he bestows on some of the most important substances.

That those he mentions constitute but a small portion of those known to his contemporaries we may infer from the fact that Pliny, though he draws from Grecian authors chiefly, many of them coeval with or older than Theophrastus, and though he mentions at least twenty minerals for each one named by Theophrastus, concludes by saying that there is no end to the names of these mineral productions, and that it is far from his intention to discourse of all.

Theophrastus distinguishes minerals according as they are capable of *effecting*, liable or not liable to *suffer* something. Thus the emerald, he observes, can impart its color to the water in which it is immersed; the Heraclean stone possesses an attractive power. Some stones may be carved, or turned in a lathe, or divided by a saw; while upon some, steel makes no impression, and others it touches with great difficulty. Sometimes the same character, as to color, hardness, and other properties, belongs to masses of considerable extent; as in the quarries of Paros, Pentelicus, and Chios. Other properties, again, belong to minerals which occur rarely and of small size; as the emerald, the sard, the carbuncle, the sapphire, and, generally, those which are engraved as seals. He also divides minerals into fusible and infusible, combustible and incombustible, and suggests various examples to illustrate these distinctions.

Pliny handles his subject in a much less philosophical way, though far more extensively than Theophrastus. In the rambling discursive manner which characterizes throughout his valuable, but ill-digested compilation, he treats of metals in two books, the 33d and 34th; of earths in the 35th; of stones in the 36th; and in the 37th and last of gems. Under which last division would have fallen, probably, most of the specimens that are found in modern cabinets.

Pliny observes that new gems, which are without name, occur from time to time. A remark that might have been made with truth at any period of the world, if by the word gem we understand, as Pliny did, a hard mineral substance, whether crystallized or not, occurring in small quantity, and fit to be engraved for ring-stones. Every year, almost, brings to light new mineral combinations, or new varieties of minerals already known. And we find that almost every province of considerable extent possesses minerals peculiar to itself. With many of these, no doubt, the ancients were unacquainted; but, on the other hand, they must have known many which modern mineralogists do not; the localities whence they were obtained having been long since exhausted, and no specimens having survived through such a lapse of time. Ancient cabinets may have contained full as great a variety as we see in modern collections; for Pliny, after specifying a great number of minerals, gives us to understand that they form but a small portion of those he could have named, and he adds a remark which might be made with equal truth of modern catalogues of minerals; that the same substance is often called by different names, in consequence of slight difference in some external characters.

Besides Theophrastus, Dioscorides, and Pliny, there are many ancient authors from whom occasionally something may be gathered respecting minerals. Historians and geographers find occasion to speak of them as occurring in countries they describe. Physicians prescribe them as remediate. Vitruvius mentions a considerable number as used in the construction or decoration of buildings. There are, moreover, works, both prose and verse, that treat expressly of minerals, and might not improperly be styled ancient, but which, falling below the age of Pliny, are beyond the limits to which our remarks will for the most part be confined.

One of the most remarkable pieces transmitted to us from antiquity, connected with our subject, is a poem consisting of near eight hundred Greek hexameters, bearing the name of Orpheus as its author, and in fact regarded not very long ago, even by learned critics, as the production of that bard so celebrated in Grecian fable; but the severity of modern criticism has disallowed its claim to such high antiquity, and left us in great uncertainty both as to its author and its date. Its chief object is to teach the mystical and healing virtues of about thirty different precious stones or mineral substances, and, due allowance being made for the poetical ornaments of thought and diction with which the subject matter is invested, it is treated in much the same way as by Pliny and other ancient naturalists. That is to say, together with some obvious characters and true properties of the mineral described, we have idle fancies and superstitious notions as to its virtues in the cure or prevention of disorders, in guarding against or healing the bite of serpents, in averting misfortune, propitiating the favor of the gods, and conciliating the good-will of men. This being a too common character of ancient works on natural science, the belles-lettres scholar, unable amid so much that is false or frivolous to discern the truths which sometimes lie hid under various disguises, often throws aside with disgust a volume in which the *scientific* reader may detect curious and useful facts.

It is proposed to make Pliny the basis of the following remarks, and, without confining ourselves very closely to his order, to combine in a sort of commentary upon the mineralogical portion of his last five books whatever in Theophrastus, Dioscorides, and others shall appear to deserve notice as connected with our subject.

But before we enter on this more methodical examination of minerals known to Pliny, at a period comparatively recent, let us consider briefly and apart those mentioned in the Bible.

OF MINERAL SUBSTANCES MENTIONED IN THE BIBLE.

The design of the sacred Scriptures was not to teach us natural science, but to make us wise unto salvation; accordingly, and in consistence with the simplicity of those early times to which the books of the Old Testament relate, we find in them few indications of any acquaintance with minerals other than six metals and various precious stones.

Besides these, indeed, the mineral substances mentioned in the Bible amount in number to no more than nine, which are marble, alabaster, lime, flint, brimstone, amber, vermilion, nitre, and salt. To these we should perhaps add two others, the one *bdellium*, mention of which twice occurs,[1] and as to the nature of which in-

[1] Gen., ii., 12; Numb., xi., 7.

terpreters are wholly at a loss;[1] the other *bitumen,* for a word in the sixteenth chapter of Genesis, translated *pitch,* and twice afterward in the same book *slime,* is thought by the learned to signify not the *vegetable* substance, but *mineral* pitch, a species of bitumen.

The only metals spoken of in Scripture as known previous to the Deluge are copper and iron, besides which we find mentioned in the Bible gold, silver, tin, and lead. Ores of two other metals appear to have been employed as pigments from the earliest times; the one a sulphuret of mercury, which furnished a native vermilion, the other a sulphuret of antimony, from which was prepared a black paint, frequently used by women in the East, even at this day, to improve the beauty of the eye by heightening its lustre and increasing its apparent size.[2]

In the short list just now given of mineral substances mentioned in the Bible there are

---

[1] The Septuagint renders it carbuncle (ἄνθραξ) in one and crystal in the other of these two passages. It is thought by some commentators to have been *pearl,* by others *beryl,* and by others again the gum-resin still called by the same name.

[2] Hence was derived one of the names (πλατυόφθαλμον) given to this pigment.

four only that require at present farther notice. These are alabaster, amber, salt, and nitre. The alabaster of the ancients was not the substance now usually designated by that name, and used to form small figures, vases, and other ornaments, which is a granular or compact gypsum: that of the ancients was more commonly a stalagmitic carbonate of lime. The name was applied both to the material and to the vessels made of it, for the purpose commonly of preserving unguents and odoriferous liquids. This is the use to which Pliny speaks of it as peculiarly adapted, and we find it mentioned in the New Testament as so employed.

As to the word which in Ezekiel, i., 4, 27, and viii., 2, is translated amber, commentators are in doubt, but seem generally inclined to think that the alloy of gold and silver hereafter to be spoken of, called by the Greeks, as amber was, ἠλέκτρον, was, rather than amber, the substance to the color of which the prophet in these passages alludes.

Table salt is a mineral with which all are familiarly acquainted, but a knowledge of the characters of that spoken of in the New Testament may throw light upon the text in which our Lord's disciples are compared to the salt of the earth, which, if it lose its savor, is cast out and trodden under foot. The salt alluded to was probably fossil salt, which occurred abundantly in various places,[1] and which containing, as such salt generally does, a large proportion of ochrey clay or other earthy matter, was liable from exposure to become insipid. Accordingly, Maundrell, in his Journey to Jerusalem, tells us that in the Valley of Salt, on the side toward Gibul, from a small precipice, formed by the continual taking away of the salt, he broke out a piece, of which the part that had been exposed to the sun, rain, and air, though it contained sparks and particles of salt, had entirely *lost its savor*, while that part next the rock still retained, as he found, its saltness.[2] Salt which had thus become insipid might be used for the purpose of repairing roads, or *cast out to be trodden under foot.*

The *nitre* repeatedly mentioned in Scripture was not our nitre, or saltpetre; but an impure carbonate of soda, procured from certain lakes in Egypt, which, as appears from recent accounts, still furnish it in great abundance. These lakes, six in number, lie west of the Del-

---

[1] See Herod., iv., 185; Maundrell and other travelers in Syria.

[2] Maundrell's Journey, p. 214.

ta of the Nile, and are called the Lakes of Natron. Hence the Greeks and Romans derived the names νίτρον, nitrum; which the Latins, as we shall hereafter see, applied, not to natron only, but to a considerable variety of substances containing more or less alkaline salts;[1] and, according to the opinion of some mineralogists, to two other wholly different compounds, chlorid of ammonium, or sal ammoniac, and nitrate of potash, or saltpetre; to which last only is the name now confined.[2]

The natron, or ancient nitre, was used for the purpose to which we continue to apply the same alkali when combined with oils, in the form of soap. The prophet Jeremiah, therefore, speaks[3] of one washing himself with nitre. And the violent action which ensues the pouring of an acid on an alkali suggested to Solomon his comparison between one "that singeth songs to a heavy heart" and vinegar upon nitre.[4] When vinegar and the salt which *we* call nitre are brought together there is no apparent discrepance between them; but pour vinegar on

---

[1] Plin., Hist. Nat., xxxi., 46. Beckmann's Hist. of Inv., iv., 525.

[2] Kidd's Min., ii., 6. Jameson's Min., ii., 316.

[3] Jeremiah, ii., 22.

[4] Prov., xxv., 20.

the nitre of Scripture, and there follows an effervescence, that shows the propriety of the royal penman's simile.[1]

OF GEMS AND PRECIOUS STONES MENTIONED IN THE BIBLE.

Let us now turn our attention to those valued productions of the mineral kingdom, which furnish the sacred writers with so many images to express beauty, magnificence, purity, solidity, and strength. As when Isaiah, foretelling the future greatness of Jerusalem, says, "Behold, I will lay thy stones with fair colors, and lay thy foundation with sapphires; and I will make thy windows of agates, and thy gates of carbuncles, and all thy borders of pleasant stones."[2] The gems mentioned here and elsewhere in the Bible, we have reason to believe, were, collectively taken, the same which the East continues still to furnish in their highest perfection; but it is impossible at this day to ascertain the species of each individual stone.

---

[1] Respecting the ancient use of nitrum in washing, see Beck., Hist. of Inv., iii., 2, 31, and Foes. in Œcon. Hippoc., p. 432. As to ancient νίτρον generally, see Coray in Hippoc. de Aere, Aqua, et Locis, vol. ii., 109, *seq.*

[2] Isaiah, liv., 12.

What in our version of the above passage is translated *agate* and by the Seventy *jasper*, Mr. Harris would call *talc*, while Bishop Lowth and Mr. Dobson would make it *ruby*. Where the name is evidently the same in the *ancient* languages and *now*, while the characters ascribed by ancients and moderns to the stone agree, we can not be in doubt; but as to others, again, a comparison of circumstances makes it evident that the ancient name, retained by us, is no longer applied to the same substance. Thus it may be inferred that the ancient chrysolite was either a deep-colored variety of the oriental topaz, or the gem by us called hyacinth; while the topaz of the ancients was the stone which we call chrysolite?[1]

Learned critics generally agree that the *diamond* was not known in the time of Moses. The word translated diamond in the description of the breastplate signifies, as is thought, a stone hard to break, or used in breaking others; a description applicable to many oriental gems. Another word (schmir) which, in Jeremiah xvii., 1, is rendered diamond, in Ezek. iii., 9, Zech. vii., 12, and Eccles. xvi., 16, is translated adamant. It signifies a very hard stone. The

[1] See afterward the article Topaz.

diamond was known at a later period among the ancients, and possessed a high value, but derived chiefly from its extreme rarity and unrivaled hardness. It was a gem of no extraordinary brilliancy nor beauty, since to display these qualities it must be cut and polished, and that art was not discovered until near the end of the 15th century.[1]

The frequent allusion made by the sacred writers to precious stones, as objects of comparison, or otherwise, may have been owing in part to their dwelling in countries near to those whence, chiefly, the most precious gems have always been obtained, and in which we may suppose them to have been less rare in those early ages than at the present day; and therefore to have furnished more familiar images of those natural qualities for which they are admired. And it is for this purpose only, of ornament and illustration, as learned commentators think, that their names are introduced. For

[1] The art of cutting and polishing the diamond was discovered in Europe by Lewis Berghem, in 1456; but some have thought it known to the artists of Hindostan and China at a very early period (Jam. Min., i., 9). Lessing, however, an acute and skillful antiquary, takes certain pretended antique cut diamonds to be sapphires. (See Brong., Tr. Elem. de Min., ii., 61, and Goquet, Orig. des Loix, ii., 111.)

although pagan antiquity ascribed various mystical virtues to certain precious stones, it is not intended that those mentioned in Scripture should be "strictly scrutinized or minutely and particularly explained as if they had each of them some precise and spiritual meaning."

It ought not to escape our notice, while upon this subject, that at the early period of the Exodus the art of polishing, setting, and even engraving precious stones was, as we learn from the description of the ephod and breastplate, already known and practiced. It is true that to engrave merely the names of the children of Israel there was not required the same degree of skill that we find afterward displayed among the Greeks, but the principles of the art and the means of execution were probably the same. And this may suggest an illustration of the difference between ancients and moderns, both as to *art* and *science*; for in regard to gems, as to other matters, it may be said that the ancients possessed less *science*, but very superior *art*. They were wholly ignorant of the chemical composition of these stones, but engraved them with a consummate taste and skill that have been rarely, if ever, equaled in modern times.

The ancients, regarding the external characters alone, and especially the color, distinguished, as was natural, by different names the sapphire, the ruby, the emerald, the topaz, and other Oriental stones, which now the mineralogist, determined by the results of analysis, classes together, as belonging all of them to a single species. One who in ancient times possessed a colorless sapphire and a polished diamond (if the ancients had been acquainted with any such) might, from their resemblance in transparency, lustre, superior weight, and hardness, have been led to place them among his treasures side by side, while the modern, in arranging his cabinet according to the true composition of minerals, would degrade the diamond from the brilliant society in which it has been used to shine, to take its place with plumbago, anthracite, and coal. To pursue this speculation farther: the artist who from a rude mass of Parian marble produced the Venus de' Medici, and he who converted a shapeless block of Lunensian marble into that form of superhuman dignity and grace, the Belvidere Apollo, knew little about the nature of the stone they chiseled, and might not easily have been brought to believe that of the solid substance upon which they wrought nearly one half the weight was an aerial acid. Ictinus, too, when he built the Parthenon, and Phidias while he adorned it with his sculpture (although the latter is said to have possessed all the science of the age he lived in), were alike ignorant of the nature of the materials they employed; but this their ignorance did not prevent the statuaries and the architect from executing works which have defied the competition of all succeeding time.

OF METALS KNOWN TO THE ANCIENTS.

From writings in which the mention of mineral substances is rare and incidental, and from which it is by inference only that we can gather any thing respecting their character or properties, let us now turn to authors who treat of them expressly, and, adopting for the most part Pliny's order, consider first the metals with which the Greeks and Romans were acquainted. These were the six already mentioned as known to the writers of the Old Testament, and, in addition to them, mercury, which is first spoken of by Aristotle and Theophrastus under the name of fluid silver (ἄργυρος χυτὸς). Its nature, however, does not seem to have been much understood even four centuries later, for Pliny distinguishes between quicksilver, *ar-*

gentum vivum, and the liquid silver, *hydrargyrus*, procured by processes which he describes from minium, or native cinnabar. This hydrargyrus he supposes to be a spurious imitation of quicksilver, and fraudulent substitute for it in various uses to which it was applied.[1]

We shall hereafter find reason for believing that the ancients had seen zinc and antimony in their metallic form; and it is probable they occasionally met with arsenic, bismuth, and other of the now known metals which occur in a native state, but confounded them with some one of the seven with which they were more familiar. They were well acquainted with zinc, arsenic, and antimony in certain of their combinations, but as reguli, or pure metals, it was by the alchemists that these three were first distinctly recognized. There has been, however, no great difference between ancients and moderns as to the state of knowledge on this head until within a period comparatively late; for although there are now, including the metallic bases of the earths, above forty metals ascertained, yet all of them, except bismuth and the ten already mentioned, have been discovered within the last hundred years.

[1] H. N.. xxxiii.. 8.

OF GOLD.

Gold, the metal which ranks highest in the estimation of mankind, was probably one of the first, if not the first, with which they were acquainted. It was found in a native state in almost every country of the ancient world, and abounded in the sands of many ancient rivers which have long since ceased to be auriferous. Such was the case with the Pactolus even in Strabo's time,[1] though at a more ancient period it had been proverbial for its golden sands. Nor are the Tagus, the Po, the Hebrus, and the Ganges any longer celebrated upon this account, as they once were; whence it seems reasonable to infer that in proportion as we go back into remote antiquity we may regard this source of supply as having been abundant. From the facility with which this gold, than which, as Pliny observes, there is none purer, was to be obtained, even in the earliest stages of society, we may account in part for the prodigious quantity on some occasions spoken of by ancient authors, as in the account given of the wealth of Solomon;[2] in the description we

[1] Strabo, Geog., p. 626.
[2] 1 Kings, vi., 20–22; x., 14–21.

find in Diodorus Siculus of the tomb of King Osymandyas,[1] and of the statues, tables, and vessels of gold dedicated by Semiramis;[2] in Herodotus's account of the wealth of Crœsus, and his presents to the oracle of Delphi, as also of the tribute paid to Darius Hystaspes by the several provinces of his empire.[3]

But this apparent abundance of gold is also to be explained in part from the fact of its having been accumulated at certain periods chiefly in some one place, as at Babylon, under Semiramis; at Jerusalem, under Solomon; at Sardis, under Crœsus; at Babylon again, under Darius, and afterward until the death of Alexander; at Alexandria, perhaps more than elsewhere, under his successors;[4] and afterward at Rome while she was at the summit of her power. Gold, too, having been used in the earlier ages almost solely for purposes of ornament, rather than as a medium of exchange, was therefore less likely to be either hoarded in the shape of money, or widely scattered through a multitude of hands. Copper having been found anciently in a native state, as well

[1] Diod. Sic., i., 49.
[2] Ibid., ii., 9.
[3] Herod., lib. i., c. 56, *seq.*, 92; lib. iii., c. 90, *seq.*
[4] See Athen. Deipn., lib. v., c. 27–36.

as gold, and sometimes in great abundance even on the surface of the ground, we hear of ancient nations using no other metals than these two. Thus we are told by Herodotus[1] that the Massagetæ use copper for their spears and other weapons of offense, gold for the ornaments of their persons; copper for the breastplates of their horses, but, for the ornaments of their reins, their bits, and their trappings, gold. Of iron and silver, he says, they make no use, nor are these metals found in their country, but copper and gold in very great abundance. He speaks of the North of Europe as especially abounding in gold,[2] and of the Scythians burying with the bodies of their kings vessels of gold, but making no use of copper or of silver.[3]

The first mention of gold and silver is where Abraham is described as "very rich in cattle, in silver, and in gold." The earliest mention of these metals as applied to any use is in relation to the ornaments which Abraham's servant presented to Rebekah, "earrings, bracelets, jewels of silver and of gold." And Pharaoh is spoken of in the same book as putting a gold chain about Joseph's neck; so that we find the most ancient uses of gold were much the

[1] Lib. i., c. 215. [2] Lib. iii., c. 116. [3] Lib. iv., c. 71.

same with those to which it still continues to be applied. But it is spoken of throughout the Bible and Homer, the most ancient books sacred and profane, as employed for a vast variety of purposes, useful as well as ornamental; and, however abundant it may have been at times, seems always to have been regarded as what Pindar styles it,[1] "a conspicuous ornament of lordly wealth." It was sometimes anciently, as it now is, applied to uses in the arts for which it was peculiarly fitted by its incorruptibility. Thus we find that it was used, even at Rome, and above three hundred years before the Christian era, for securing in their places artificial teeth—a law of the twelve tables making an exception as regards such gold, and permitting it to be buried together with the dead.[2]

Among the uses to which gold was anciently applied, it may deserve mention that a tissue was sometimes made of it without admixture of any other substance. Pliny says[3] he had seen Agrippina, wife of Claudius, seated at his side during the exhibition of a naval combat, clothed in a robe of gold woven without other

[1] 1 Olym., 4. [2] Cic., de Leg., ii., 24.

[3] H. N., xxxiii., 19.

material. Dion Cassius speaks of the same occasion, when Agrippina was, according to him, arrayed in a cloak of gold (χλαμύδι διαχρυσίῳ ἐκοσμεῖτο). Lampridius, in his Life of Heliogabalus, mentions a tunic of the same kind worn by him,[1] and Sidonius Apollinaris alludes to the mode of weaving such cloth of gold.[2]

Though it is probable the auriferous sands of rivers and alluvial plains furnished anciently, as perhaps they do still, the greatest quantity of gold, yet there were mines worked at a very early period. Cadmus is said to have opened the first mine of copper and gold in Mount Pangæum;[3] the same region from which, eleven hundred years later, Philip of Macedon derived the treasures which enabled him to prosecute successfully his ambitious schemes. The Pharaohs drew great quantities of gold from mines situated on the borders of Egypt and Ethiopia, between the Nile and the Red Sea. The processes by which it was obtained, as described by Diodorus,[4] are to be seen represented in paintings executed in the reign of Osirtasen I.,

[1] Aldi edit., fo. 112. [2] Carm. xxii., v. 119.

[3] Strabo, p. 998.

[4] Diod. Sic., lib. iii., c. 12, 13, 14, and the original source in Photii Biblioth., c. 1339.

seventeen centuries before the Christian era. It appears that mines of copper and of gold were worked in Siberia at a very remote period by some people unacquainted with the use of iron tools.[1] These mines, which are on the southern and eastern borders of the Ural Mountains, and have been examined of late years by Gmelin, Lepechin, and Pallas, were probably among the sources whence those Scythians of whom Herodotus makes mention drew their gold.

There were in Thasos, and other Greek islands, very ancient gold mines, originally opened by the Phœnicians. Herodotus tells us he had himself seen these Phœnician mines of Thasos, and that a great mountain had been overturned there in searching for the metal.[2]

Spain also contained rich mines of gold; and an ancient city of its Atlantic coast, Tartessus, is thought by Bochart to have been the Tarshish of Scripture.[3] Strabo declares[4] that no country in the world produced gold, silver, copper, and iron in so great abundance, and of so good a quality, as that part of Spain called Turdita-

[1] Jacob's Hist. of the Precious Metals, c. 2.
[2] Lib. vi., c. 47. [3] Geog. Sacra, iii., 7, col. 170.
[4] Geog., p. 146.

nia, and its neighborhood; and Pliny observes[1] that the barren mountains of this country, which yielded nothing else, were rich in gold. The same author speaks of gold mines in the territory of Vercellæ, in Italy, in which to employ more than five thousand men was expressly prohibited by law.[2] Diodorus Siculus mentions Arabia as producing the finest native gold (χρυσὸς ἄπυρος), in pieces about the size of a chestnut, and of so bright a color that artists used it in setting the most precious stones to form beautiful ornaments.[3]

OF SILVER.

Silver and gold, as they are generally found associated in varying proportions, were anciently, in many cases, gotten from the same mines. Pliny observes that silver is derived from mines only; that it was found in almost all the Roman provinces, but the best in Spain; and was, like gold, obtained from a barren soil and mountains. Gaul must be excepted from the provinces that furnished silver; for, according to Diodorus, none was found there; but the sands of its rivers yielded much gold.[4] The mines

---

[1] H. N., xxxiii., 21.
[2] Ibid.
[3] Lib. ii., c. 50.
[4] Lib. v., c. 27.

opened in Spain by Hannibal, certain of which had yielded him three hundred pounds of silver daily, were in Pliny's time not yet exhausted, and the mountain that contained them had been excavated to the distance of a mile and a half.[1] This country would appear to have been among the principal sources of this metal; and, perhaps, hence chiefly even the East may have drawn its supplies. But many other countries furnished silver besides Spain; Egypt in enormous quantity, if the account of Diodorus (i., 49) is to be received; and while upon this subject it would not be right to omit all mention of that "fountain of silver, treasure of the earth," which Æschylus says[2] the Athenians possessed. The mines of Laurium, to which he alludes, were probably more productive about this time than at any period before or since. They do not appear to have been at any period very profitable, as compared with those of other countries, and though they continued for many centuries to employ great numbers of men, they, during a great portion of the time, scarcely defrayed the expenses of working them. Strabo speaks of them as originally valuable, but in his time exhausted.[3]

[1] H. N., xxxiii., 31.

[2] Æschyl., Pers., v. 238.

[3] Geog., p. 399, where see Casaubon. These mines are

Silver and gold are spoken of together on occasion of the presents, before alluded to, brought by Abraham's servant to Rebekah, "jewels of silver and jewels of gold;" but a mention of silver, as applied to use, occurs in the preceding chapter, where Ephron values his field at four hundred skekels of silver, which are weighed to him by Abraham, and styled *current with the merchants,* that is, probably, of the due fineness. From this passage, as also from that in which Joseph's brothers are said to have sold him for twenty pieces of silver;[1] from their taking with them silver on both occasions of their going into Egypt to buy corn; and from Joseph's accumulating all the silver of Egypt and of Canaan in exchange for corn,[2] we may infer that at a very early period silver, though not coined, was a usual medium of exchange. This was a use to which gold does not appear to have

---

mentioned by Herodotus, Thucydides, Pausanias, and others; and Xenophon wrote his treatise *περί προσόδων*, or on the revenues of Attica, to encourage the Athenians to work them. They seem to have been sufficiently valuable to be coveted by Philip, as from various passages in Demosthenes appears.

[1] Gen., xxxvii., 28. Here the Septuagint has *εἴκοσι χρυσῶν*, twenty pieces of gold.

[2] Gen., xxxvii., 14. In all these passages the Septuagint has *silver;* but in our Bible the word used is *money.*

been so soon applied, though in other respects the two metals were employed for like purposes. We find drinking-vessels, statues, idols, altars, ornaments of temples, houses, and the person, made sometimes of one, sometimes of the other metal; and, in the same ornament, or article of furniture, frequently the two combined.[1] Mirrors in Pliny's time were commonly of silver. He says the best had been those manufactured at Brundisium, of a mixture of tin and copper,[2] until even the female slaves began to use them made of silver. It seems they were sometimes of this metal as early as when Plautus wrote;[3] and Seneca speaks of them in his time as being made "of the full length of the body, of silver and of gold, carved and adorned with precious stones."[4]

OF ELECTRUM.

Electrum, a native alloy of gold and silver, is a compound of which the ancients sometimes speak as though it were a simple substance, for the reason, probably, that they were unable

---

[1] Hom., Od., δ., v. 125; η., v. 89. [2] H. N., xxxiii., 45.

[3] Plaut., Mostell., i., 3, 111.

[4] "Totis paria corporibus, auro argentoque cœlata, gemmis adornata."—*Sen.*, *Nat. Quæst.*, i., 17.

readily to separate the metals of which it was composed.

Pliny, having observed that all gold contains more or less of silver, adds that when the silver is in the proportion of one fifth the alloy is called *electrum*, which, he says, is also artificially compounded, silver being added to gold in the required proportion.[1] He understands Homer as meaning this alloy where ἤλεκτρον is mentioned in the description of the palace of Menelaus;[2] but Eustathius thinks that in this passage amber is the substance meant.[3]

Klaproth has applied the name electrum to argentiferous native gold, to which, as we have seen, it anciently belonged, when the silver was in the above-mentioned proportion of one fifth.[4]

---

[1] H. N., ix., 108. [2] Hom., Od., δ., v. 73.

[3] Dr. Clarke (Trav., vol. viii., p. 231) describes a coin of Rhescuporis, a Thracian prince, made of this alloy, and takes occasion to point out the error into which several authors have fallen as to passages in ancient writers, in which the alloy and not amber is intended.

[4] Sophocles, though he styles the Pactolus εὔχρυσον (Phil., v. 393), elsewhere (Antig., v. 1037) speaks of the *electrum* of Sardis together with the gold of India, from which it might be inferred that the gold of the Pactolus (on which Sardis stood) was argentiferous, contained a sufficient proportion of silver to cause it to be denominated ἤλεκτρον.

OF IRON AND COPPER.

From the two more precious metals, gold and silver, we come next to two which possess a much greater value in use than in exchange; and which, in the times we now treat of, stood in such close relation to each other that we shall find it convenient to consider them, in some measure, under the same head.

Iron, as well as copper, was in use before the Deluge; for Tubal-cain, we are told, was "an instructor of every artificer in brass and iron."[1] It is not probable that a knowledge of such useful auxiliaries, once acquired by man, would ever have been lost. Æschylus must have regarded the use of iron as exceeding ancient; for he makes Prometheus boast of having taught it to mankind.[2]

The art of working iron, therefore, and the use of weapons and other things made of it, appears to have been known among the Egyptians at a very early period; and Moses was well acquainted with this metal, though neither by him in the construction of the tabernacle in the wilderness, nor by Solomon afterward in the building of the temple, does it appear to have been employed.

[1] Gen., iv., 22. [2] Prom. V., v. 502.

Moses compares the deliverance of the Israelites from Egyptian bondage to their being "brought forth out of the iron furnace."[1] He uses the terms iron and brass indifferently, in a figurative sense, as emblematic of something stern and hard: "I will make your heaven as iron, and your earth as brass;"[2] and elsewhere, "Thy heaven that is over thy head shall be brass, and the earth that is under thee shall be iron."[3] He mentions "a yoke of iron," also, in a figurative sense. We are told that the bedstead of Og, King of Bashan, "was a bedstead of iron."[4] Mines of iron are spoken of;[5] and it appears that hostile weapons, and tools for cutting stone, were sometimes made of iron.[6] However, during times of high antiquity, brass or copper was much more used than iron, especially among nations that had made little progress in the arts. Hesiod speaks of iron as, during the brazen age, unknown: "Their weapons, dwellings, tools, were all of brass."[7]

---

[1] Deut., iv., 20. [2] Levit., xxvi., 19.
[3] Deut., xxviii., 23. [4] Deut., iii., 11.
[5] Deut., viii., 9. [6] Num., xxxv., 16; Deut., xxvii., 5.
[7] Hesiod, ἔργ. καὶ ἡμ., v. 150. Lucretius, too, says, lib. v., v. 1286,

"Et prior æris erat quam ferri cognitus usus."

It may perhaps be worthy of remark that the sleep of

The reason of this is obvious. Iron is not found in a native or metallic state, except in aerolites, and the skill in metallurgy was often insufficient to reduce it from its ore, and work it; but native copper is occasionally found in almost every country, and sometimes in large masses even on the surface of the earth; and this metal and its alloys may be worked with much greater ease.[1] Accordingly, uncivilized tribes have been discovered, on the northwest coast of America and elsewhere, making use of copper implements; and it is with good reason that Werner conjectures it was the first metal worked by man. Axes and other instruments of bronze have been found in tombs of the ancient inhabitants of Mexico and Peru; and from the latter country Humboldt took a chisel, which Vauquelin ascertained to consist of .94 of copper and .06 of tin. This alloy was so well forged that it had a specific gravity of 8.815, a maximum of density which, according to the experiments of Mr. Briche, a chemist does not obtain but by a mixture of 16 parts of tin with 100 of copper.

---

death, which Homer (Il., xi., 241) calls χάλκεον ὕπνον, Virgil, writing at a period when iron was more familiarly known, styles ferreus somnus (Æn., x., 745).

[1] See Goguet, Orig. des Loix, etc., t. i., p. 141.

Brass was a very indefinite term among the ancients, the simple metal copper, and all the different compounds into which it entered as a principal ingredient, being comprehended under the same name, χαλκὸς, æs. When, therefore, in Scripture and in ancient authors generally, brass is spoken of, we are seldom to understand the alloy now designated by that name, although that is sometimes meant; but the brass tempered for edge-tools, or formed into warlike weapons, was generally a compound of copper and of tin; and the proportion between the two metals seems to have been, in many cases, nearly that just now stated as affording the maximum of density. The bronze of an ancient Grecian helmet was found, on analysis, to be copper alloyed with 18.5 per cent. of tin.[1] An ancient dagger, analyzed by Hielm, was found to contain about five parts of copper and one of tin; and other analyses of ancient weapons, found in countries widely distant from each other, have afforded much the same result.[2] The same metals, combined in

[1] Phil. Trans., 1826, part ii., p. 55.

[2] See Ure's Chem. Dict., art. Copper. The nearly identical result obtained in such a diversity of cases favors the belief that this ancient brass was a native alloy, procured

various proportions, are still applied to many important uses; and sometimes to mould brazen cannon, those more formidable instruments of modern war. Dr. Clarke gives the results of several analyses of ancient bronze, and observes that the constituent metals, copper and tin, are in every case nearly the same; but he makes the proportion of tin to be only 12 per cent.[1]

It is probable that, in remote antiquity, copper was oftener employed than any alloy of it. Indeed, there still remain both warlike weapons and artisans' tools made of the simple metal. From Homer's saying that the spear of

---

from long since exhausted sources of an ore in which, as in the bell-metal ore of Cornwall, tin and copper were united, and in the proportions requisite to give the compound its greatest density and hardness; and this may be that brass of which Pliny speaks (xxxiv., 2) as in his time no longer found, the mines that once furnished it having become exhausted. If such a native alloy was found in Etruria, as some have thought, that circumstance might help to account for the early civilization and the wealth of that country. One of the earliest trading voyages that we read of—that of Mentes in the Odyssey—is represented as having for its object to exchange iron for bronze at a port of Italy, Temese or Brundisium. A native alloy of bronze, or a mineral furnishing such, is said to have been found in Mexico, where instruments of the above composition have been obtained.

[1] Travels, vol. vi., p. 505; vol. v., p. 292.

Iphidamas did not pierce the girdle of Agamemnon, but that "its point was turned as though it had been lead when it struck upon the silver,"[1] we may fairly infer that the spear-head was copper rather than any hard alloy of it. In the ancient mines of Siberia, which have lately been examined, hammers and wedges of copper have been found.[2] There is reason, therefore, to suppose that the words χαλκὸς and *æs* should be rendered by the word *copper* much oftener than they are.

Pliny distinguishes *copper* as æs Cyprium—brass of Cyprus—in which island, he informs us, it was first discovered; but he gives the name *æs* to the alloys of copper, not with zinc only, but with gold, silver, tin, and lead, with all which metals it was mixed, and in different proportions, according to the color or other qualities required in the compound, or the uses to which it was to be applied. Among the rest, Pliny specifies three varieties of Corinthian brass, more precious, he remarks, than silver, nay, almost preferred even to gold itself.

Some of the most valued productions of ancient art, were in brass of Delos and Ægina,

---

[1] Hom., Il., λ., v. 236.

[2] Jacobs's Hist. of the Prec. Met., p. 27.

those two islands having been celebrated for their manufacture of a material well suited for the statuary. Myron used the brass of Delos, and Polycletus that of Ægina.[1]

Pliny gives for statuary brass a recipe, according to which there is added to the melted copper one third part of old brass, worn and polished by use (æs collectaneum), with twelve and a half per centum of an alloy called plumbum argentarium, which consisted of equal parts of lead and tin.[2]

Aristotle informs us that the Mossynœcians had anciently prepared a brass of a pale color and superior lustre, mixing it, not with tin, but with a certain earth found among them.[3] This earth was, no doubt, the cadmia of Pliny, our calamine, an ore of zinc; though we have reason to believe that Pliny has comprehended under the term cadmia an ore of copper abounding in zinc, from which was procured the brass called *orichalcum*, or mountain brass, the kind most highly valued; but the mines of *orichalcum* becoming exhausted, as we find that in Pliny's time they were,[4] an imitation of the na-

[1] H. N., xxxiv., 4, 5. [2] Ibid., xxxiv., 48.

[3] Arist., Op., v. i., p. 1155, B. "Cadmia terra, quæ in æs conjicitur ut fiat orichalcum."—*Festus, de Verb. Sig.*

[4] H. N., xxxiv., 2.

tive alloy was produced by mixing with copper the ore of zinc called calamine.[1] Keferstein supposes,[2] but perverts the language of Aristotle to favor his idea, that the Mossynœcian brass was white copper, an alloy of copper and arsenic. That the ancients, however, were acquainted with and made use of a white copper is undoubted, for in the Herculanean Museum there are pateræ and other articles made of such alloy;[3] and perhaps the *æs candidum* mentioned by Pliny as used together with copper and tin in making mirrors[4] may have been obtained from copper ores that abounded in arsenic, which now also enters into the composition of speculum metal. The *æs candidum* spoken of as one variety of the Corinthian brass seems to have owed its color to the large quantity of silver it contained. When Virgil speaks of a cuirass, "auro squalentem alboque orichalco," the epithet, as Cerda and others think, indicates

---

[1] This was our brass (Watson's Chem. Ess., iv., 85), and should be distinguished from the more ancient bronze, "the old chemical compound of copper and tin;" for of this also Dr. Clarke supposes some ancient native alloy to have existed.—*Clarke's Trav.*, vol. vii., pref., p. xxiii.

[2] Phil. Mag., vol. lxiii., p. 126.

[3] Winck., Mon. Ant. Ined., vol. ii., p. 172.

[4] H. N., xxxiv., 48.

merely the splendor of the brass, as Hesiod and Apollonius Rhodius speak ὀρειχάλκοιο φαεινοῦ. That orichalcum was of the color of our brass may be inferred from the case which Cicero supposes of a man ignorantly selling gold by mistake for it.[1]

The uses to which copper and its alloys were applied are far too numerous to admit of being specified. The mechanics' tools, implements of agriculture, weapons and defensive armor, useful and ornamental articles of household furniture, coined money, statues, in some cases doors, columns, and even roofs were made of brass.[2] With the advance of society and the improvement of the arts of life in modern times, we find the use of iron continually extended, and the purposes to which it is applied daily more and more diversified. In proportion as we go back into antiquity we find its use more rare, and its place supplied by other metals; thus nails and bolts were made of brass,[3] and the

---

1 Cic., de Off., iii., 23.

2 Plin., H. N., xxxiv., 7. Respecting the abundance of brass in the earlier times of Rome, and the probable sources of it, there are some remarks in Niebuhr's Hist. of Rome, vol. i., p. 451.

3 Dioscor., v., 88; Athenæi Deipnos, v., 40; Gell's Argolis, p. 29.

anchors of the Phœnicians were of lead.[1] The Romans, however, used iron, for purposes both of war and peace, at least as early as the sixth century before the Christian era; for Pliny speaks of a stipulation in the treaty made with them by Porsenna which prohibited their use of iron except in the cultivation of the soil.[2]

In the Trojan age, and in Homer's time, iron was well known; and the poet employs, as Moses does, the names of iron and brass in both a literal and a figurative sense, as though the two metals were applied indifferently to the same purposes, whether of war or peace, and regarded as possessing in like manner the properties of hardness and tenacity. Iron, however, was used much more sparingly than brass, being, for the reasons no doubt that have been stated, much the rarer of the two. Accordingly we find that a rudely-cast round mass of iron (σόλος αυτοχόωνος), which had been used by Eëtion as a quoit, is represented by Achilles, when he offers it among the prizes at Patroclus's funeral, as "a five years' provision of iron for one

[1] Diod. Sic., v., 35.

[2] H. N., xxxiv., 39. If Etruria derived advantage from its trade in bronze, this stipulation may have been merely with a view to commercial profit.

who cultivates extensive and rich fields, so that his shepherd or his plowman, when they need iron, may, without going to the city for it, satisfy their wants."[1] Hence we learn that it was used, though in very small quantity, for implements of agriculture. This passage, certainly the most remarkable one in which iron is spoken of in Homer, is sometimes referred to as though it were the only one, but most incorrectly, as we shall perceive. It is mentioned as an article of commerce where Mentes tells Telemachus that he has iron which he is going to exchange at Temese for copper.[2] Axes were sometimes made of iron. A chariot-builder is spoken of as felling a tree *αἴθωνι σιδήρῳ*.[3] Those axes through which the suitors were to shoot, often as they are spoken of, are always called iron (*σίδηρος*) or polished iron (*πολιός σίδηρος*).[4] Weapons of war were sometimes made of it. The iron weapon is said to invite men, that is, to tempt them, to violence.[5] Antilochus holds the hands of Achilles, lest he should cut his own throat *σιδήρῳ*.[6] An "iron

---

[1] Il., ψ., 826. [2] Od., α., 184. [3] Il., δ., 485.
[4] See Hom., Od., δ., 587; φ., 3, 81, 97, 114, 127, 328; ω., 167. [5] Od., π., 294; τ., 13. [6] Il., σ., 34.

arrow-head" is mentioned,[1] and Areithous breaks the phalanxes with "an iron club."[2]

We find both brass and iron mentioned in the same passage or description. Iris, preparing the chariot of Juno, places the wheels of brass on the iron axle (Il., ε., 723). The gates of hell are of iron and the threshold is of brass (Il., η., 15).

The tempering of iron by plunging it into water is alluded to.[3] It is repeatedly styled "much-labored" (*πολύκμητος*), and is classed with gold and brass as a thing of value. Among the treasures collected by Ulysses are mentioned "brass, and gold, and much-labored iron."[4] Achilles enumerates as part of his booty, "gold, and red brass, and bright iron" (*πολιόν τε σίδηρον*).[5] Adrastus speaks of his father's treasures, which Menelaus should receive as ransom for his life, "brass, and gold, and much-labored iron;"[6] and the sons of Antimachus offer the same ransom to Agamemnon.[7]

The scarcity and value of iron in remote ages is inferred by Mr. Jacobs from discoveries

[1] Il., δ., 123. [2] Il., η., 140. [3] Od., ι., 393.
[4] Od., ξ., 324; φ., 10. [5] Il., ι., 366.
[6] Il., ζ., 48. [7] Il., λ., 133.

made of late years in the ancient Scandinavian tumuli. "There are swords, daggers, and knives, the blades of which are of gold, while an edge of iron is formed for the purpose of cutting. Some of the tools and weapons are formed principally of copper, with edges of iron; and in many of the implements the profuse application of copper and of gold, when contrasted with the parsimony evident in the expenditure of iron, seems to prove that, at the unknown period, and among the unknown people who raised the tumuli which antiquarian research has lately explored, gold as well as copper were much more abundant products than iron."[1]

The high value set upon iron in the reign of Alyattes, the father of Crœsus, about 600 years A.C.N., and the slight acquaintance at that period with the working of it, may be inferred from his dedication at Delphi of a wrought-iron base for a bowl, the work of Glaucus of Chios, who first invented the welding of iron, for that seems to be the only meaning we can give to the words of Herodotus (i., 25), *σιδήρου κόλλησιν*.

But to return to Homer for a moment: he repeatedly makes use of the term *σίδηρος*, *iron*,

[1] Jacobs's Hist. of the Prec. Met., p. 10.

and the adjectives derived from it, in a figurative sense. The *iron* tumult is said to reach the *brazen* heaven,[1] and the insolence and violence of the suitors to reach the *iron* heaven.[2] The iron strength of fire is spoken of,[3] and a resemblance in strength αἴθωνι σιδήρῳ.[4] The bodies of the Greeks are said not to be of stone or iron, so as to resist the piercing brass,[5] and Euryclea compares herself to "firm stone or iron."[6] Telemachus is told that his father will not remain much longer absent, "even though iron chains should hold him;"[7] and he speaks of his father as having perished, "although he had an iron heart within him."[8] Ulysses is told that he is insensible to the sufferings of his companions, because he himself knows not fatigue—"that he is altogether formed of iron;"[9] and Hecuba, thinking Priam insensible to the danger he is about to brave, declares "he has an iron heart."[10] Ulysses' eyes are said to remain as unmoved as if they were of horn or iron.[11]

Hesiod speaks of bronze more frequently than of iron;[12] but, like Homer, he uses in a

---

[1] Il., ρ., 424. [2] Od., ο., 328. [3] Il., ψ., 177.
[4] Il., υ., 372. [5] Il., δ., 510. [6] Od., τ., 494.
[7] Od., α., 204. [8] Od., δ., 293. [9] Od., μ., 280.
[10] Il., ω., 205. [11] Od., τ., 211.
[12] Theog., v. 722, 724, 726, 732, 733, 749, 750.

figurative sense epithets derived from either metal, as in Theog., v. 764:

"Τοῦ δὲ σιδηρέη μὲν κραδίη, χάλκεον δέ οἱ ἦτορ."

While speaking of ancient uses of iron, it may be worth while to notice a curious passage found among the fragments of Ctesias, from which we might perhaps infer that the ancients, 400 years before the Christian era, had some obscure notion of the use of lightning-rods. Ctesias speaks of a fountain of India, from the bottom of which was obtained a kind of iron, of which he had two swords, and which was of such a nature that, being implanted in the ground, it averted clouds, and hail, and lightnings—πηγνύμενος ἐν τῇ γῇ, νέφους, καὶ χαλάζης, καὶ πρηστήρων ἐστιν ἀποτρόπαιος.[1]

The Greeks of a much later age than Homer's marked the distinction between steel and iron by calling the former *στόμωμα*, the word *στομόω* signifying to give a temper or keen edge to a cutting instrument.

The ancients sometimes made steel by a process that has been employed even in modern times. The crude iron, purified by repeat-

[1] Photii Bibliotheca, p. 144, or Herod., ex edit. Gronovii, p. 656.

ed melting, and deprived of a portion of its carbon by being kept long in a state of fusion, was converted into steel. This at least seems to be the rationale of the mode described by Aristotle, who says[1] the iron was purified from the scoria by melting, and when it had been treated thus several times, and became pure, was changed to steel (στόμωμα). Ancient steel is thought to have been made in some cases, as German steel is said to be, directly from cast iron, merely by forging the crude metal to a certain point.[2] A mode of making steel in use among the Celtiberi is thus described by Diodorus Siculus:[3] "They bury in the earth," says he, "forged plates of iron, and leave them till in length of time the rust has consumed the feebler parts of the metal, and left the firmer, of which they make excellent swords and other weapons, such that neither shield, nor helmet, nor bone are able to resist them." Plutarch[4] and Suidas[5] also speak of this same excellent steel; and a similar process is said to be still used in Japan, where the iron

[1] Vol. i., p. 590.
[2] Parke's Chem. Essays, vol. i., p. 10.
[3] Lib. v., c. 33. [4] Plut., de Garrulitate, vol. ii., p. 510.
[5] Word μάχαιρα, vol. ii., p. 510.

is buried in marshy ground till a great part of it is consumed by rust, when it is taken up and forged, and again buried for eight or ten years, when what remains is found converted into a sort of steel, of which they make plowshares and other tools and weapons. The Japanese sabres are said to be incomparable, and of such temper that they will easily cut through a nail without injuring their edge.[1]

Some portion of the iron-work of one of the London bridges having been taken up from a considerable depth, where it had been buried for a great length of time, was found to furnish a very superior material for knives and razors. Cast iron immersed in sea-water for a long time is converted into a sort of plumbago, and the burial of wrought iron in earth may produce in it a change of a similar nature, but less in degree.

But, besides other ancient modes of preparing steel, there can be little doubt that it was in some instances obtained directly from such ores as are now sometimes called steel ores, simply by the process used in making bar-iron; and the virtues which Pliny[2] ascribes to the

---

[1] Beck., Hist. of Inv., iv., 242, 243.

[2] H. N., xxxiv., 41.

waters of certain places, as Bilbilis in Spain and Comum in Italy, may have belonged rather to the ores which were there usually employed.

But, whatever may have been the way in which the ancients made their steel, this at least is certain, that they possessed instruments of such a temper as to carve porphyry and other stones that resist the tools of modern artists.[1] It is not certain, however, that those instruments were of steel. Modern experiments have shown that bronze of the ancient composition may, by cooling it slowly, be rendered as hard as steel, and at the same time less brittle. Wilkinson (Anc. Egypt., vol. i., 320; vol. ii., 344) describes Egyptian knives and daggers of bronze which, after the lapse of several thousand years, retain their flexibility and their spring, having an elasticity which resembles that of steel.

Steel was sometimes called *χάλυψ*, chalybs, because obtained, of an excellent quality, from the country of the Chalybes. Eudoxus, as cited by Stephanus, De Urbibus (*χάλυβες*, p. 714), says, *ἐκ δὲ τῆς χαλύβων χώρας ὁ σίδηρος ὁ*

[1] Respecting the mode and the difficulty of working porphyry, see Winckelmann, Storia delle Arti del Disegno, vol. i., p. 88; vol. ii., p. 15.

περὶ τὰ στομώματα ἐπαινούμενος ἐξάγεται—from the country of the Chalybes is derived the iron which is most approved for the steeled parts of instruments. This is supposed to have been "the northern iron and the steel" mentioned in Jeremiah, xv., 12. The Indian steel (σίδηρος Ἰνδικὸς καὶ στόμωμα) mentioned by the author of the Periplus[1] was probably of the kind still brought from India under the name of wootz; and the *ferrum candidum* of which Quintus Curtius says[2] the Indians presented to Alexander a hundred talents may have been the same, for wootz when polished has a silvery lustre;[3] and that kind which Pliny styles *Sericum,* and to which he assigns the palm, was perhaps no other.[4] The Parthian steel ranks next with Pliny, and these two kinds only "mera acie temperantur."

Daimachus, a writer contemporary with Alexander the Great, speaks of four different kinds of steel, and the purposes to which they were severally suited. "Of steels (τῶν στομωμάτων) there is the Chalybdic, the Synopic, the Lydian, and the Lacedæmonian. The Cha-

[1] See Plin., Exercitationes, p. 763, b F.
[2] Lib. ix., c. 8. [3] Beck., Hist. of Inv., iv., 248.
[4] H. N., xxxiv., 41.

lybdic is best for carpenters' tools; the Lacedæmonian for files, and drills, and gravers, and stone chisels; the Lydian also is suited for files, and for knives, and razors, and rasps."[1]

What Aristotle, Daimachus, and Eudoxus, in the passages that have been cited, call *στόμωμα*, and Plutarch calls *σιδήρου στόμωμα καὶ ἀκμὴν*,[2] is by Latin writers called *acies, ferri acies.* This, in the barbarous latinity of a later period, came to be called *acierium*, and hence the French *acier*. The Greek and Latin terms seem to indicate that steel was employed, as in many cases it still is, merely for the edge, "ad indurandam aciem," of cutting instruments, and for such part of other instruments as needed its harder temper, so that *στόμωμα* did not so much denote steel itself as the steeled part of the instrument.[3]

OF LEAD AND TIN.

Lead and Tin are metals which we have the best reason for treating of under the same head, since the ancients frequently confounded them,

1 See Stephanus, De Urbibus, word *Lacedæmon;* and Fabricii Bib. Græc., vol. ii., p. 588.

2 Vol. ii., p. 693, A.

3 Beck., Hist. of Inv., iv., 236.

and, however strange may appear such confusion in regard to metals so plainly distinguished by their properties as these, their names, nevertheless, in Hebrew, Arabic, Greek, and Latin are often indifferently used.[1]

The Greeks, when they would distinguish the two metals, called tin κασσίτερος and lead μόλιβδος, but, as the French at this day call both pewter and tin *etain*, so did the Greeks comprehend under the name κασσίτερος various alloys of tin with lead or some other metal; and some such compound Homer is supposed to mean when he speaks of tin (κασσίτερος) used in the fabrication or ornament of various parts of armor.[2]

The Romans distinguished lead (plumbum) into black and white.[3] The latter, plumbum album, was the more precious, Pliny says, being what by the Greeks was called κασσίτερος.

Plumbum album[4] is sometimes called stan-

---

[1] Salm., De Hom. Hyl. Iatr., p. 234.

[2] Mention of κασσίτερος occurs ten times in the Iliad; once as used about a chariot, but on every other occasion about breastplate, shield, or greaves.

[3] Plin., H. N., xxxiv., 47.

[4] What *we* term *white lead*, the ψιμμύθιον of the Greeks and cerussa of the Romans, was made anciently by processes

num, while on other occasions the latter is spoken of as something different, in which case it may have been an alloy of tin and lead, or, as Beckmann thinks, of silver and lead; or it may have been designated by a different name merely because obtained from a different place, from an ore of different appearance, or by some different process, since any one of these, we know, was anciently sufficient ground of distinction between substances that were in fact identical. If any reliance could be placed on Pliny's accuracy in a matter of this kind, we might infer from what he says of the mode in which stannum was obtained[1] that the ancients were acquainted with an argentiferous galena containing also tin. Beckmann, however, in his examination of this passage, says that lead is seldom found without, but that tin, perhaps, has never been found with silver. He admits that the passage in question can not be fully understood with any explanation, yet he thinks it proves to conviction that the stannum of the ancients was not tin, but a mixture of silver and lead, called in the German smelting-houses *werk*. It is from stannum, however, that are

that are in use even at the present day. (Dioscor., v., 103; Plin., xxxiv., 54.)

[1] H. N., xxxiv., 47.

derived the names *etain* and *tin.* He supposes the oldest κασσίτερος to have been nothing else than the stannum of the Romans. Aristotle, however, ascribes to what he calls τὸν κασσίτερον τὸν Κελτίκον a property which is characteristic of the metal tin.[1]

Pliny treats as a fable the account of tin being brought from islands in the Atlantic Sea, the only localities with which he seems acquainted being Gallicia and Lusitania, where it was found, he says, on the surface of the earth, in grains of a black color, and detected only by its weight. It is now found in Gallicia, in veins traversing granite and mica slate.

Herodotus too professes his ignorance even of the existence of the islands called Cassiterides, whence tin was brought;[2] but Strabo describes them,[3] ten in number, as possessing mines of tin and lead, which, together with skins, they exchanged for earthen-ware, salt, and brazen vessels with the merchants who visited them, the Phœnicians alone having at first carried on this trade from Cadiz. These Cassiterides, which Strabo speaks of as distinct

---

1 Beckmann, Hist. of Inventions, iv., 21; Aristotle, Auscult. Mirabil., vol. i., p. 1154.

2 Herod., iii., 115.

3 Geog., p. 75.

from Britain, were what we now call the Scilly Islands, and probably received the metals they traded in from the opposite coast of Cornwall. Diodorus Siculus regards them as belonging to Spain rather than to Britain, and Dionysius Periegetes speaks of them as inhabited by Spaniards. The former author, having observed that tin occurs in many places in Spain, not on the surface, as some historians have reported, but dug from the earth and smelted as gold and silver are, adds that there are many mines of it in islands, which thence derive their name, Cassiterides, lying over against the coast of Spain; and that much was brought from the island of Britain, also, through Gaul to Marseilles and Narbonne.[1] The latter author, having in view perhaps their western position only, calls them the Hesperides, the source of tin.

"Νήσους θ' Ἑσπερίδας, τόθι κασσιτέροιο γενέθλη."[2]

Lead was obtained in abundance from Spain, Gaul, and Britain,[3] and applied to much the same uses as in modern times. Mixed with tin in equal proportion it constituted plumbum argentarium, and with one third part of tin terti-

[1] Diod. Sic., lib. v., c. 38. [2] Dion. Perieg., v. 563.
[3] Plin., H. N., xxxiv., 49.

arium. These were used in the composition of statuary bronze, and the latter of them for the solder of leaden pipes, in making which pipes a vast quantity of lead must have been employed, considering the mode in which ancient cities were supplied with water, and the common use of baths.[1] It was also used, as copper now is, in sheathing ships, and, as we shall see, in preparing the common pigments white and red lead. It or its ore was indispensable in the refining of silver, as Pliny tells us,[2] and Theognis speaks of it more than five centuries earlier as used in refining gold,[3] as do likewise Diodorus Siculus[4] and others.

But, among the uses to which plumbum album and stannum were applied, well deserves to be noticed the coating, or tinning, as we term it, of copper vessels (which, Pliny tells us, rendered the taste more agreeable, and guarded against the poisonous effects of verdigris) and the covering in like manner ornamental articles of copper, so that they could scarcely be distinguished from silver. For this manufacture were famous the Bituriges, the modern Bourges; and another town of Gaul, Alesia, the

[1] See Vitruv., viii., 6.
[2] H. N., xxxiii., 31.
[3] Theog., Γνωμ., v. 1101.
[4] Lib. iii., c. 13.

modern Alise, afterward employed silver for a like purpose, the plating of harness and carriages of various kinds.[1]

The ore from which the ancients procured their lead was that which now furnishes almost all the lead of commerce—the sulphuret, termed by us galena—a name which Pliny also uses,[2] but as synonymous with molybdena, which is described by Dioscorides[3] and himself as an argentiferous ore of lead. This latter name, molybdena, has been adopted by us to designate a new and wholly different metal.

OF MERCURY.

Mercury has already been spoken of as known to the ancients at least as early as Aristotle's time. Vitruvius describes a useful application of its property of forming with gold an amalgam, and says that silver and copper can not be rightly gilt without it;[4] and the mode of doing this Pliny has described.[5]

It seems proper to notice here again, in con-

---

[1] H. N., xxxiv., 48.
[2] Ib., xxxiv., 53.
[3] Dioscor., v., 100.
[4] Vitruv., vii., 8.
[5] H. N., xxxiii., 20, 42. Consult also, respecting the method of gilding made use of by the ancients, Winckelmann, Stor. delle Arti del Disegno, vol. ii., p. 29.

nection with the subject of metals, terra cadmia or calamine, an ore of zinc already spoken of, a substance with which the ancients were well acquainted, though they are commonly supposed not to have known zinc itself, except as combined with copper in the form of brass. But a passage in Strabo authorizes the belief that they also knew this metal in a separate state. The geographer says[1] that near Andeira, a town of Troas, is found a stone which being burned becomes iron, and distills false silver (ἀποστάζει ψευδάργυρον) when heated in a furnace together with a certain earth, which, receiving the addition of copper, forms the alloy that some call brass (ὀρείχαλκον). He adds respecting this *false silver,* which was probably our zinc, that it occurs also near the Tmolus. Stephanus states the same thing in somewhat clearer words, and refers to both Theopompus and Strabo as authorities.[2]

This earth, which is said to derive its name cadmia from Cadmus, son of Agenor,[3] who first introduced at Thebes the making of brass,[4] is

---

1 Strabo, p. 610.
2 Steph., De Urbibus, word *Andeira.*
3 See Hardouin on Pliny, vol. ix., p. 195.
4 Hygini Fab., 272.

spoken of by Aristotle, in a passage before referred to,[1] but by no name; and Theophrastus also alludes to it without naming it; for, after having spoken of the characters and properties of stones, when he comes next to consider the different kinds of earths, he mentions, as a very peculiar one, that "which mingled with copper is capable not only of melting and blending with it, but has the extraordinary property of changing and improving its color."[2] Pliny repeatedly speaks of cadmia, but it is evident that he does not always mean one and the same thing. Cadmia seems to have signified with him not only our calamine, but a copper ore which contained zinc; and the same name was extended to what the Germans call *offenbruch*, furnace calamine, which in melting ores that contain zinc or in making brass falls to the bottom of the furnace, and contains more or less of calcined zinc.[3] There is little reason to doubt that the varieties of cadmia which Dioscorides calls ὀστρακίτις, πλακώδης, and βοτρυίτις were also furnace calamine. Pliny certainly speaks of them as such, and describes together with them a fourth variety, which he calls capni-

[1] See *ante*, p. 46. [2] De Lapid., c. 84.

[3] Beck., Hist. of Inv., vol. iii., p. 70, 72.

tis.[1] The contrary opinion of Saumaise is founded on what Beckmann seems justly to regard as an annotation that has been admitted into the text of Dioscorides.

The pompholyx and spodos of Dioscorides[2] and Pliny[3] were oxides of zinc, more or less pure, produced by combustion of the calamine in the process of making brass. There was no essential difference between the two, but spodos was the heavier and of a darker color, being less completely oxidized, or less perfectly sublimated, and mixed with impurities from the walls and pavement of the furnace. Pompholyx was the white oxide called flowers of zinc or lana philosophica, and is compared by Dioscorides also to carded wool.[4]

OF ANTIMONY.

A sulphuret of Antimony, called by the Greeks *στίμμι* and by the Romans stibium, was from the earliest times and still is used in the East for tinging black the hair and eyebrows, the eyelashes, and edges of the lids, this last application being with a view to increase the apparent size of the eye, wherefore *στίμμι* was

[1] H. N., xxxiv., 22.
[2] Dioscor., v., 85.
[3] H. N., xxxiv., 33.
[4] Dioscor., v., 85.

sometimes called πλατυόφθαλμον. The use of this cosmetic is twice spoken of in Scripture, at least according to the Seventy, for our version has not in either passage[1] specified the kind or color of the paint employed. Pliny's description of stibium[2] does not suit in all respects the common sulphuret of antimony; but this mineral may have been found then more frequently associated, as it now sometimes is, with the white oxide or with the nickeliferous sulphuret, to either of which Pliny's description of it as "candida nitensque" might with propriety be applied. In preparing it as a paint, it is, according to Dioscorides,[3] to be inclosed in a lump of dough, and that buried in coals until reduced to a cinder: extinguished with milk and wine, it is to be again placed upon coals and blown until ignition, but if burned longer it becomes lead (μολιβδοῦται). Pliny directs cow-dung to be used in place of dough, and varies so entirely from the recipe of Dioscorides that it is evident he had some other authority before him, yet he too recommends as especially necessary to observe moderation in burning it, lest it should be converted into lead (ne

[1] 2 Kings, ix., 30; Ezekiel, xxiii., 40.

[2] H. N., xxxiii., 33.

[3] Dioscor., v., 99.

plumbum fiat).[1] The fair inference from these passages appears to be that the ancients occasionally saw antimony also reduced to its metallic state, but, as in the case of zinc,[2] confounded it in a loose and careless way with another metal better known to them.

METALLIC SUBSTANCES, EARTHS, ETC., EMPLOYED AS PIGMENTS BY THE ANCIENTS.

The most famous ancient painters—Apelles, Echion, Melanthius, and Nichomachus—used only four colors, the white melinum, the Attic sil, the sinopis of Pontus, and atramentum.[3]

Colors are divided by Pliny into bright and dull (floridi et austeri). The florid or bright pigments are those with which the employer furnishes the artist, as minium, armenium, cinnabari, chrysocolla, indicum, purpurissum. Other colors are austere or dull. Pigments are again divided into native—as sinopis, rubrica, paraetonium, melinum, eretria, auripigmentum—and factitious, of which also divers kinds are specified.[4]

In considering these substances separately, it will not be necesary to observe exactly the order in which they are here named.

---

[1] H. N., xxxiii., 34.
[2] See *ante*, p. 66.
[3] Plin., H. N., xxxv., 7.
[4] H. N., xxxv., 12.

Κιννάβαρι. Theoph., c. 103. Ἄμμιον. Dioscor., lib. v., c. 109, 110. *Minium.* Vitruv., lib. vii., c. 8, 9; Plin., lib. xxxiii., c. 40.

This pigment was, as we learn from Theophrastus,[1] either native or factitious. The native came from Spain and Colchis, rough and stony; the factitious from a single place not far from Ephesus. The process of preparing it from a bright scarlet sand there found is described by Theophrastus as the invention, about ninety years before his time, of one Callias, an Athenian, who was led by the shining appearance of this sand to expect gold from it—a hope which its unusual weight may have contributed to encourage. This Ephesian *κιννάβαρι* of Theophrastus, which by Vitruvius and Pliny is called minium, we are perfectly sure, from what Vitruvius says, was sulphuret of mercury. That, however, Dioscorides calls *ἄμμιον*, and appropriates the name *κιννάβαρι* to a substance brought from Libya, which some, he says, thought to be dragon's blood. Arrian

[1] Here, and in all that follows respecting ancient pigments, a general reference to Theophrastus, Dioscorides, Vitruvius, or Pliny is intended as a reference to that portion of their works cited at the beginning of each article respectively.

and Pliny speak of an Indian cinnabar, which the former correctly enough describes as the inspissated tear-drop, as it were, of trees, while the latter regards it as the mingled and concreted blood of a dragon and a dying elephant crushing the serpent by its weight.[1] From this strange notion has been derived the vulgar name of dragon's blood, sometimes applied to reddle, and sometimes to a dark-red resin imported from the East Indies, the substance which Arrian intended to describe.

Minium was a pigment highly valued, and even held sacred among the ancients. With it were painted the statues of the gods and the persons of those who triumphed.[2] The titles of books also, at a somewhat later period, as well as the initial and more important letters, were decorated and rendered more conspicuous by means of it, and those who were employed in doing this were, from their use of minium,

---

[1] Arrian, Peripl. Mar. Eryth., p. 159; and Plin., H. N., xxxiii., 7, edit. Var.

[2] Plin., H. N., xxxiii., 36; Winckelmann, Stor. delle A. del Dis., vol. i., p. 16; Pausan., p. 115, and Kuhn's note thereon. With this practice of painting red the persons of victors might be compared that of our North American Indians.

called *miniatores*. The word miniature is probably derived from this practice of illuminating the initial letters in manuscripts, these *miniaturæ* being originally monochromata, for which, whenever painted, the ancients commonly made use of minium.[1]

In Pliny's time minium, though found elsewhere, was brought from scarce any place but Spain, whence native cinnabar to the amount of about ten thousand pounds' weight annually was sent under seal to Rome, where the exclusive privilege of manufacturing vermilion seems to have been enjoyed by a company, which, not content with its fair profits, derived a farther gain from various adulterations of the genuine pigment. The fraudulent substitute for minium spoken of by Pliny as found in almost all silver and lead mines, and prepared by burning thoroughly the stone associated with the silver ore, was probably red-lead.[2]

---

[1] Winck., Stor. delle A. del Dis., vol. ii., p. 60; Plin., xxxiii., 39. Our word *rubric* has a like origin with that suggested for *miniature*, the red oxyde of iron, called *rubrica*, having been used as a substitute for minium to illuminate initial letters. Hence Juvenal (Sat. xiv., v. 192) calls the laws of the ancients *rubras*, and Persius (Sat. v., v. 90) speaks of Massurii rubrica.

[2] H. N., xxxiii., 40.

Μίλτος. Theoph., c. 91–95; Dioscor., lib. v., c. 111, 112. *Rubrica.* Vitruv., lib. vii., c. 7; Plin., lib. xxxv., c. 13, 14.

This more common red pigment among the ancients was the ochrey red oxyde of iron. The best, Theophrastus says, was that of Cea, but the Lemnian and the Sinopic too were celebrated. The latter, found in Cappadocia, and carried down to Sinope, derived thence its name, which was often applied to the like substance, from whatever country it might have been obtained. This red ochre was found varying considerably as to compactness, fineness, and intensity of color, and passing sometimes into reddle and bole. The terra Lemnia or sigillata was bole,[1] and is sometimes distinguished from the rubrica of the same island, which is the kind of rubrica preferred by Dioscorides and Pliny. The terra Lemnia (which, differing little from rubrica, is regarded by Pliny as the same) was never sold unsealed, and was therefore called σφραγὶς, terra sigillata, the sphragide of Jameson. It was stamped before the time of Dioscorides[2] with the figure of a goat; afterward, in Galen's time, with the image of Diana; and has been of later years with

[1] Cleav. Min., p. 473.

[2] Dioscor., v., 113.

the seal of the Turkish empire. It was formerly much used in medicine as an astringent.

From Egypt and Carthage was obtained an inferior kind of *μίλτος* or rubrica used by carpenters,[1] who for coloring their line made use of red more commonly, but sometimes black,[2] and not of white, as we do. Accordingly, a white line being one that made no mark, to use a white line, *λευκὴ στάθμη*, was a proverbial expression for *to act without discernment.* This more common or cheaper red ochre was sometimes made by burning yellow ochre, which was the discovery, Theophrastus says, of one Cydias, who, where a certain inn had been consumed by fire, saw half-burned yellow ochre converted into red. Both Theophrastus and Vitruvius[3] describe the mode of preparing this red pigment, which the latter writer and Pliny call *usta,* though Pliny comprehends under the same name, *usta,* another common red pigment, red-lead, discovered, as the last mentioned was, accidentally, from jars containing white-lead (cerussa) having been exposed to the heat of a conflagration in the Piræus.[4]

---

[1] Dioscor., v., 113.

[2] Eustath. in Hom. Od., ε., ν., 245.

[3] Vitruv., vii., 11.

[4] H. N., xxxv., 20.

Σανδαράκη. Theoph., c. 71; Dioscor., v., 122. *Sandaraca.* Plin., xxxv., 22; Vitruv., vii., 7.

This red pigment was the sulphuret of arsenic, called realgar. An adulterate kind of it was made, Pliny says, of calcined white-lead; that is, the red-lead he had just before described under the name *usta* was substituted for realgar. But Vitruvius prefers to the native sandaraca this substitute, which he designates by no other name, simply saying that cerussa is by the heat of a furnace converted into sandaraca. A mixture of sandaraca with rubrica in equal proportions made the paint called sandix, and a farther addition of sinopis produced that called syricum.[1] Strabo speaks[2] of a mine of Sandaraca at Pompeiopolis, in Paphlagonia, in which, because of the dangerous exhalations from the mineral, none other were employed but slaves who had been sold on account of crime.

Purpurissum, a red pigment, which, because of its great cost, was furnished by employers,[3] was composed of creta argentaria dyed in a decoction of hysginum,[4] or of madder, as Dutch

---

1 Plin., H. N., xxxv., 23, 24.

2 Strabo, p. 562.

3 Plin., H. N., xxxiv., 12.

4 A crimson dye made from an insect found on the plant called ὕσγη.

pink is made by combining with clay or marl the coloring matter of woad or that obtained from the berries of the yellow-leaved buckthorn (Rhamnus infectorius). The creta spoken of by Pliny and by Vitruvius as the basis of purpurissum, armenium, cœrulium, factitious chrysocolla, and other pigments was generally an argillaceous earth. It may, however, sometimes have been chalk, if indeed the Greeks and Romans were acquainted with any chalk adapted to such use. It is very probable, moreover, that it was in some cases a white talc, like that called chalk of Briançon, which when reduced to a fine powder and colored with the flowers of the Carthamus tinctorius[1] is still used, as the purpurissum of the Romans was,[2] to give an artificial bloom to the complexion.

Ὤχρα. Theoph., c. 115; Dioscor., v., 108.
Sil. Vitruv., vii., 7; Plin., xxxiii., 56.

Yellow ochre, which the Greeks called simply ὤχρα, and the Romans *sil,* appears to have been the principal yellow pigment of the ancients. Pliny specifies three varieties: the Attic, which was best; the marmorosum, which may have been what we call stone ochre; and

---

[1] Bastard saffron.

[2] Plaut., Mostel., i., 3, 104; Trucul., ii., 2, 35.

the Syricum, of a dull color, named from the island Syros, as may have been the red paint, also, called by the same name. Sil was found in many places, Vitruvius observes, but the Attic, which used to be the best, was no longer to be obtained, because the veins of it, which occurred in the silver mines of Attica, were no longer now explored. It is described by Dioscorides as light, smooth, free from stone, friable, and of a full bright yellow; and probably any ochre possessed of these qualities continued in Pliny's time to be called Attic. Theophrastus[1] speaks of painters using ochre (ὤχρα) instead of orpiment (ἀῤῥενίκον), because, however they might seem to differ, there was, in fact, no difference of color.

Polygnotus and Mycon were the first who painted with sil. The Attic, which they used, and that called lucidum, which came from Gaul, were afterward employed for the lights of pictures, and the Syricum, the Achæan, and the Lydian for the obscurer parts.

Ἀῤῥενίκον. Theoph., c. 71, 89, 90. Ἀρσενικόν. Dioscorides, v., 121. *Auripigmentum.* Vitruv., vii., 7; Plin., xxxiii., 22. *Arsenicum.* Plin., xxxiv., 56.

The yellow sulphuret of arsenic derives its

[1] De Lapid., c. 90.

name, orpiment, from one of its Latin appellations, auripigmentum, and it was so called, perhaps, not merely from its golden color and the use to which it was applied, but because the ancients thought it really contained that metal. Pliny mentions, among other modes of obtaining gold, that of making it from orpiment, and says that Caius (Caligula) ordered a great quantity of that substance to be reduced, "and certainly obtained excellent gold, but in such small proportion as to lose by an experiment which was not afterward repeated." Although no great reliance can be placed on this account, we are not of necessity to regard it as a fable, for the mass experimented on may have contained, as it is said this mineral sometimes does, a small portion of gold.[1]

The Greek name ἀρσενικὸν (masculine) was given to this sulphuret because of the potent qualities it was discovered to possess—qualities which the present arsenic of the shops, a white oxyde of the metal, exhibits in a more in-

[1] Hill says he had a fine specimen of orpiment from the mines of Gosselaer, with veins of native sandarach running across it, which was brought to him as a gold ore, and which, he thought, really contained a small quantity of that metal. —*Hill's Trans. of Theoph.*, p. 176.

tense degree. The ancients were well aware of the kindred nature of this yellow sulphuret and the red sulphuret before mentioned, called sandaraca. Theophrastus always mentions them together; Dioscorides observes that they are found in the same mine; Celsus that they possess in all respects the same virtue;[1] and Pliny that they are of the same substance. The two were then, in fact, as now, found occasionally intermixed. "There is a third kind," says Pliny, "in which the golden color is mingled with the red."

Theophrastus and Pliny speak of both these minerals as found in gold and silver mines, and of orpiment Pliny says that it is dug in Syria, for the use of painters, near the surface of the ground, of a golden color, friable like the specular stone (lapidum specularium modo), that is, the laminæ were easily separable "like those of mica."[2] Vitruvius mentions Pontus as a locality, and Dioscorides names Mysia as the country whence the best was brought, that of Pontus holding the second rank.

Massicot, the yellow oxyde of lead, which Sir H. Davy thinks was used as a pigment by the ancients,[3] may have been comprehended by

[1] Cels., v., 5.

[2] Cleav. Min., p. 679.

[3] See Ure's Chem. Dict., article *Paints*.

them under the name chrysitis, one of the three varieties of litharge (λιθάργυρος) described by Dioscorides,[1] and by Pliny, who has copied him. Its name (χρυσῖτις) was no doubt derived from its being, as Dioscorides describes it, yellow and shining (ξανθὴ καὶ στίλβουσα), and in so far resembling gold. Though Dioscorides speaks only of its medicinal properties, yet, since the seven substances which form the subjects of his seven next succeeding chapters were all pigments, we may reasonably suppose chrysitis also to have been such.

Χρυσοκόλλα. Theoph., c. 46; Dioscor., v., 104. *Chrysocolla.* Vitruv., vii., 9; Plin., xxxiii., c. 26, 27, 28.

This was a green pigment much spoken of among the ancients, and which has been with the moderns a subject of much uncertainty and doubt. It is probable that under one and the same denomination were comprehended malachite, copper-green, and various green earths. The name applied to those several substances properly belonged to a factitious compound, described by Pliny,[2] used in soldering gold. In this chrysocolla (gold glue), "which the goldsmiths claim as theirs" (aurifices sibi vindi-

[1] Dioscor., v., 102.

[2] H. N., ix., 29.

cant), the most important ingredient, perhaps, was the subcarbonate of soda (nitrum).[1] The best chrysocolla, Pliny tells us, was procured from copper mines, the next best from silver mines, the third quality from gold mines, and that least valued from lead mines. His description of it, as a substance which from a liquid flowing through mineral veins had become indurated, agrees well with the undulated appearance often presented by the surface of compact malachite, and its color, too, is that of malachite, being most approved, he says, when it resembled that of luxuriant corn while in the tender blade. When used in medicine it excites vomiting, he says, and Dioscorides, too, observes that it is one of the substances "which promote vomiting and can cause death." This author states that the best was brought from Armenia, the second kind from Macedonia, and the third from Cyprus.

Theophrastus speaks of "native cyanos containing chrysocolla,"[2] or, in modern terms, of the blue carbonate of copper containing, as it

---

[1] Some have been so far mistaken as to suppose the chrysocolla of Pliny to be the biborate of soda (borax) now used by jewelers in soldering gold. (See Jameson's Mineralogy, ii., 350.)

[2] De Lapid., c. 70.

often does, the green; and Aristotle also mentions that the two occur together in the island Demonesus.[1] Kidd agrees with others in thinking the chrysocolla of Theophrastus to be malachite, but supposes the Greek term to have in this case a passive signification, and to imply that the substance was "set in gold," it being sometimes polished as an ornamental stone.[2] The impure ores of copper, called by some mineralogists copper-green, are by others, as Brongniart, Aikin, and Phillips, denominated chrysocolla, and it is most probable that some of them were by the ancients comprehended under that name, as were also some of those green earths now known as mountain-green, or green earth of Saxony, from its being there obtained, as it is likewise at Kernhausen in Hungary, near Verona in Italy, and in various parts of the United States.[3] Vitruvius speaks of a green earth, fit for the use of painters, found in many places, but the best at Smyrna.[4] Celsus speaks of viride Alexandrinum,[5] and Pliny of viride Appianum.[6] These earths differ consid-

---

[1] Arist., De Mirab. Auscult., vol. i., p. 1154.
[2] Kidd's Min., ii., 120.
[3] Cleav. Min., p. 445.
[4] Vitruv., vii., 7.
[5] Cels., v., c. 26, § 23.
[6] H. N., xxxv., 29.

erably from place to place in their composition, but do not often contain copper, to which metal they were once supposed to owe their color. That the name chrysocolla was very loosely used is evident; for Pliny has described one kind obtained by the evaporation during June and July of waters suffered to flow through metallic veins during the preceding winter and until the month of June; and, if his text here were not corrupt, he would probably be found to describe another factitious kind in much the same way that Vitruvius does, who says that those who can not use chrysocolla, because of its high price, dye a blue pigment (cœruleum) with the herb called lutum (which produced a yellow color), and thus obtain a lively green.

From Dr. Ure's analysis of an ancient green pigment brought by Mr. Wilkinson from Thebes, it was found to consist of blue glass in powder, yellow ochre, and particles of a colorless glass, which served to give it a brighter hue.[1]

Ἰός. Theoph., c. 102; Dioscor., v., 91. *Ærugo.* Vitruv., vii., 12; Plin., xxxvi., 26.

Verdigris was among the ancients, as it is at this day, a common green pigment, and may be

[1] Wilk., Anc. Egyp., vol. iii., p. 302.

considered as belonging to our subject, seeing that Dioscorides and Pliny specify several varieties of native ἴος or ærugo, classing with it in this case what we may suppose to have been green carbonate instead of acetate of copper; as, for example, "the efflorescence upon stones which contained copper," and what was "scraped from the stone out of which copper was melted."

Various modes of making verdigris are described by Theophrastus, Dioscorides, and Pliny, which agree in principle, and some of them even as to their details, with processes now used. Among the various adulterations of it, that made with the sulphate of iron (atramentum sutorium) was, as we learn from Pliny, the one best calculated to deceive, and the mode of detecting it suggested by him is deserving notice. It was to rub the counterfeit ærugo on papyrus steeped with the gall-nut, which immediately thereon turned black.

Κυανός. Theoph., c. 98; Dioscor., v., 106. *Cœruleum.* Vitruv., vii., 11; Plin., xxxiii., 57.

Under these names were comprehended by the ancients several blue pigments differing widely in their origin and nature. Theophrastus, having observed that cyanus (κυανὸς) is either native (αὐτοφυής) or factitious (σκευαστὸς), as in Egypt, specifies three kinds, the Egyptian, the Scythian, and the Cyprian, of which the two last named appear to have been native, and in that case were probably the blue carbonate of copper. What the Egyptian cyanus was has been a subject of much doubt. Some travelers have thought the blue which may still be recognized in the paintings of Thebes to be ultramarine;[1] and Beckmann infers that the blue seen on mummies, having lost little or nothing of its brightness, must be either ultramarine or cobalt, but with the latter he thinks it can be shown that the Egyptians were unacquainted.[2] Modern analyses, however, have shown his mistake, and that the blue on ancient pottery, both Egyptian and Chinese, was sometimes derived from cobalt.

Theophrastus speaks of the artificial cyanus as prepared by means of fire (πεπυρωμένος), which agrees well with the idea of its being ultramarine; but so does it also with the account which Vitruvius gives of this matter—an account which, it seems, might have settled

[1] See Norden's Travels, vol. ii., p. 51.

[2] Hist. of Inv., ii., 360.

all doubts, had it received the attention it deserves. The composition of cœruleum, he says, was first invented at Alexandria,[1] and a manufactory of it afterward established at Puteoli by Vestorius. It was made by reducing sand mingled with pure nitrum (subcarbonate of soda) to a fine powder, which, being sprinkled with coarse copper filings, was formed into balls and dried. The balls were then arranged in earthen vessels, and placed in a furnace, until by its heat they were converted into cœruleum. Now this is precisely the cœruleum prepared by Sir H. Davy,[2] who found that 15 parts of carbonate of soda, 20 parts of opaque flint powdered, and 3 parts by weight of copper filings, strongly heated together for two hours, yielded a compound which when powdered produced a fine deep sky-blue, which he regarded as the same azure employed in some ancient paintings.[3] But there are chemists

[1] There must be some inaccuracy here, if Vitruvius has in view, as is most probable, the Egyptian azure spoken of by Theophrastus; for he, writing about the time of the foundation of Alexandria, says that "those who write of the Egyptian kings mention the king who first made artificial azure in imitation of the native," c. 98.

[2] See Paris's Life of Sir H. Davy, vol. ii., p. 48, 299.

[3] See Ure's Chem. Dict., article *Paints*, Sir H. Davy

who regard this preparation, called "cendres bleus," as not a very permanent pigment, and therefore doubt if any of the blues now remaining in Egypt were so prepared.

As to the native azure, there seems little reason to doubt that it was generally, if not always, blue carbonate of copper.[1] Pliny observes[2] that it is found in gold and silver mines, and Theophrastus speaks of it as occurring together with chrysocolla, that is, with malachite. This blue carbonate was at one time much used in Persia as a pigment,[3] and is so used in some countries to this day.[4] Pliny mentions as other varieties of cœruleum, the Puteolanum, the Vestorianum, and the Hispaniense. The first two we may infer from what Vitruvius says to have been the same with each other and with the Egyptian, and the last, which seems, like the others, to have been factitious, probably differed from them only as to the place from which it came. The cœruleum styled Indicum, which Pliny speaks of as not long before

---

seems to differ from Beckmann in supposing the ancients to have been acquainted with cobalt.

[1] Beck., Hist. of Inv., vol. ii., p. 325, 329.

[2] H. N., xxxiii., 56. [3] Beck., Hist. of Inv., ii., 317.

[4] Cleav. Min., p. 568.

his time introduced, was probably indigo.[1] We find from Dioscorides[2] and Vitruvius[3] that it came from India; and the former, though he does not describe it as a vegetable substance, supposes it to be a concretion about Indian reeds. Pliny, however, mentions, as a test by which the true cœruleum Indicum might be distinguished, that, placed upon a coal, it exhibited a brilliant purple flame, "flammam excellentis purpuræ"—a property of cœruleum which could not belong to such mineral substances as have heretofore been spoken of under that denomination, but characteristic of indigo, which is sublimed by heat, producing a splendid blue or purple vapor.

Pliny says a fraudulent imitation of cœruleum was produced by a decoction of dry violets strained through a linen cloth upon creta Eretria; and he speaks obscurely of still another factitious kind, which in Vitruvius[4] we find described more clearly as an imitation of the Indian azure obtained by dyeing creta Selinusia or argentaria with woad, the ἴσατις of the

---

[1] See Schneider's Comment. on Vitruv., lib. vii., c. 9; and Beck., Hist. of Inv., vol. iv., p. 103, 105.

[2] Lib. v., 107.

[3] Lib. vii., 9.

[4] Vitruv., vii., 14.

Greeks, called vitrum or glastum by the Romans.

Ἀρμένιον. Dioscor., v., 105. *Armenium.* Vitruv., vii., 9; Pliny, xxxv., 28.

This blue pigment, called after the country whence it came, is described by Pliny as differing from cœruleum in that it has a whitish tinge. The kind which by Dioscorides is esteemed the best appears to have been an earth, for he requires it to be smooth, friable, and free from stone. It was probably the same mineral he speaks of in the next chapter under the name *cyanus*, only that it was in an earthy form, and came from Armenia instead of Cyprus. In its medicinal effects it resembled chrysocolla, but was less efficacious. Besides this earthy blue pigment there was also a stone called lapis Armenius, and this is thought to have been some quartzy or calcareous substance penetrated by the blue carbonate of copper. Jameson says the Armenian stone of the ancients was a limestone impregnated with earthy azure copper ore, and in which copper and iron pyrites were sometimes disseminated.

Among the blue pigments made use of by the ancients Jameson would include the earthy phosphate of iron, "for a substance answering to blue iron earth is mentioned by Pliny as being collected in the marshes of Egypt, and ground, and washed, and used as a pigment;" and there are peat-bogs in New Jersey where a similar earthy phosphate of iron is found.[1]

Besides the several white earths employed as pigments by the ancients, they used also white-lead, called *ψιμύθιον*, cerussa; and the mention of it here may be allowable, since Pliny speaks[2] as if there were a native kind that had at one time been in use, although when he wrote it was artificially prepared. The processes described by Theophrastus,[3] Dioscorides,[4] Vitruvius,[5] and himself, do not differ essentially from those now used. The substance spoken of by Pliny as a native cerussa, found at Smyrna on the farm of Theodotus, appears to have been that greenish earth mentioned by Vitruvius[6] as occurring in many places, but the best near Smyrna, and called by the Greeks *θεοδότιον*, from the name of the person, Theodotus, upon whose farm it was first discovered. From the fact that this greenish earth was regarded as a

---

[1] See Renwick's Outlines of Geology.

[2] H. N., xxxv., 19.

[3] Theoph., c. 101.

[4] Dioscor., v., 103.

[5] Vitruv., vii., 12.

[6] Vitruv., vii., 7.

sort of ceruse, we might infer that the ceruse of the ancients was not always of a very pure white.

OF EARTHS USED IN MEDICINE OR AS PIGMENTS.

Theophrastus, Dioscorides, Vitruvius, Pliny, and other ancient authors speak of various earths used in medicine chiefly, or as pigments, and the second named of these writers specifies certain properties of these earths that were possessed in common by them all.[1] Indeed, with the exception of *ampelite*, they seem to have resembled each other greatly in their supposed medicinal virtues, as well as in their external characters. Thus, the Selinusian had the same effects with the Chian,[2] and this the same with the Samian.[3] The Pnigitis resembled in color the Eretrian, and was sold in place of it by some, and it possessed the same virtues with the Cimolian.[4] These earths were in most cases argillaceous, and by the term creta was generally meant some whitish clay, such as potter's clay, pipe-clay, or fuller's earth. It was

---

[1] Dioscor., v., 170. [2] Ib., v., 175. [3] Ib., v., 174.
[4] Ib., v., 177; Plin., xxxv., 56.

sometimes, however, applied to a calcareous marl, of which kind probably were the "candida fossitia creta," used in Gaul as a manure,[1] and the "creta pulvis," which Palladius recommends for the same purpose;[2] and we have reason to believe that where such magnesian earths occurred as those varieties of talc and steatite which we call French and Spanish chalk, the ancients comprehended them also under the name *creta.* That creta was clay we might infer from the uses to which it was applied; but we are besides expressly told it was so by the writers "De Re Rustica" and others. "Creta, quam argillam dicimus," says Palladius.[3] "Creta, qua utuntur figuli, quamque nonnulli argillam vocant," says Columella.[4] These writers speak repeatedly of "creta figularis"[5]—"creta qua fiunt amphoræ."[6] Celsus, too, speaks of "creta figularis,"[7] and Vitruvius of "vas ex creta factum, non coctum."[8]

One of the most celebrated of those earths was the creta Cimolia, much used, both in med-

---

1 Varro, i., 7, 8.
2 Feb., xxv., 22.
3 Pallad., i., 34, 3.
4 Colum., iii., 11, 9.
5 Col., iii., 11, 9; vi., 17, 6; viii., 2, 3; Veg., iii., 4.
6 Col., xii., 4, 5.
7 Cels., i., 3.
8 Vitruv., viii., 1, 5.

icine and as a fuller's earth for scouring and cleansing woolen cloths. Both uses of it seem to have been more ancient at least than the age of Hippocrates, for he more than once prescribes[1] detersive earth (σμηκτρίδα γῆν), which Erotianus and Galen explain to mean Cimolia (τὴν Κιμωλίαν). This earth, which anciently gave celebrity to Cimolus,[2] still forms beds in that island, now called Argentiera, and in Milo. It was in modern times first recognized by Tournefort,[3] and is now called by mineralogists cimolite, argile cimolithe. It is grayish white, but by exposure to the air becomes reddish, whence probably is derived Pliny's distinction of it into two kinds, "candidum, et ad purpurissum inclinans."[4] It possesses the detergent property of fuller's earth, but is less unctuous to the touch than that variety of clay. Mr. Hawkins found that the terra cimolia which he procured at Argentiera cleansed woolens as well as the best fuller's earth. The inhabitants of the island still use it, not only for this purpose, but as a substitute for soap in wash-

[1] Hippocr., Op., p. 667, 1; p. 884, E.

[2] "Hinc humilem Myconen, cretosaque rura Cimoli."—Ov., Met., vii., 463.

[3] Jam. Min., i., 478.

[4] H. N., xxxv., 57.

ing linens; though, from what Tournefort observes, it seems badly suited to such use, inasmuch as the grains of sand which it contains are apt to wear away the linen.[1] The name cimola was applied to other earths besides that of the island Cimolus, from their possessing in greater or less degree its detersive properties. That of Sardinia Pliny styles "vilissima omnium cimolæ generum"—the cheapest kind of cimolite. This creta Sarda was, however, used in the first place to cleanse garments that were not dyed, which were then fumigated with sulphur, and finally scoured with cimolia.[2] We perceive, therefore, that when "cretata vestis," "cretata ambitio candidati," are spoken of we are not, as some have done, to fancy that the gown of the candidate was whitened with a sprinkling of powdered *chalk*, but that it was cleansed and brightened by the fuller's art.

The first of the earths enumerated by Pliny, collyrium, is still represented in our cabinets by the kollyrite, argile kollyrite; and of the last one in his arrangement and in that of Dioscorides, ampelitis, we retain the name, under

[1] Brongn., Min., i., p. 519; Tournefort's Voyage, vol. i., p. 113, Eng. trans.

[2] Plin., H. N., xxxv., 57.

which some mineralogists have included aluminous and graphic slate.[1]

The ampelitis of the ancients appears to have been a bituminous shale, and is called by Strabo "the bituminous earth ampelitis."[2] This earth was found near Seleucia in Syria, was black, and resembled small pine charcoal, and when rubbed to powder would dissolve in a little oil poured upon it. Its name was derived from its being used to anoint the vine (ἀμπέλος), and preserve it from the attack of worms[3]—a use to which it is said that bitumen has been recently applied. The Cilician earth spoken of by Theophrastus,[4] which when boiled became viscid, and with which, instead of birdlime, vines were anointed to guard them against insects, was probably something of a like kind.

Collyrium was so called from its supposed virtue as an external application to sore eyes,[5] such remedies being called collyria. It was one of the two more remarkable varieties of

---

[1] Brongniart (Min., i., 561) agrees with Messrs. Romé-de-Lille and Haüy in thinking ampelitis to have been the same with the minerals which he describes under the names ampelite alumineux and ampelite graphique.

[2] Strab., p. 316.

[3] Dioscor., v., 181.

[4] De Lapid., c. 85.

[5] Plin., H. N., xxxv., 53.

Samian earth, the other of which, called aster, resembled probably the *ochra Attica vel asterace* of Celsus.[1]

The earths had their several names from the places where they respectively occurred. The Selinusian was sometimes called argentaria, because used to polish silver. The earth spoken of in connection with Selinusian, and called annularia, which was stained with woad to produce an imitation of indicum,[2] may have been the same with the annulare mentioned afterward,[3] which was so called because made of clay colored with common glass ring stones. This at least, strange as it is, appears to be the only sense we can extract from Pliny's words, the meaning of which Beckmann acknowledges he had not been able to discover.[4] The same author inclines to think that annularia received its name from its use in sealing—a purpose to which certain kinds of earth were anciently applied.[5] Another conjecture finds annularia in the clay-stones which occur in curious circular

---

[1] Celsus, v., 14, and Milligan in Celsum, p. 201.

[2] Plin., H. N., xxxv., 27. [3] Ib., xxxv., 30.

[4] Hist. of Inv., iv., 106.

[5] See Herod., ii., 38; Cic. pro Flacco, c. 16; Beck., Hist. of Inv., i., 208.

nodules, and sometimes in the form of rings. Pnigitis was probably obtained from the neighborhood of a village called Pnigeus, on the coast of Egypt, not far from which was a promontory of white earth called the White Coast (Λεύκη 'Ακτή).[1] Nearly related to Pnigitis was Parætonium, one of the earths most commonly made use of as a white pigment,[2] and which derived its name from a town in Egypt, where it was obtained, not far from Alexandria. It was a heavy, tough clay, of a fine white color, and in the mass often contained minute shells and other impurities, owing probably to its having been rolled by the waves upon the beach, after being washed in a purer state out of sea-beaten cliffs in which strata of it were contained. It was procured from Cyrene also, and from Crete,[3] and, Hardouin says, is found in Saxony, and known by the same name.[4]

Melinum, or terra Melia, was so called from the island Melos (Milo), where it was obtained. It was of an ashy white, resembling the ash-colored Eretrian earth, and was useful in paint-

[1] Strabo, p. 799.
[2] Vitruv., 77; Plin., H. N., xxxv., 18.
[3] Plin., H. N., xxxv., 18.
[4] Hard. in Plin., vol. ix., p. 414.

ing to give permanence to colors.[1] One of the several varieties of Samian earth resembled it, but, being unctuous, stiff, and slippery, was not employed by painters.[2] The Melian was applied to the same uses in medicine as the Eretrian earth.

Theophrastus, speaking of factitious pigments, adds that there are three or four varieties of native earths, which from their superior usefulness deserve mention; and he specifies the Melian, the Cimolian, the Samian, and the Tymphaican, which last he farther characterizes as gypsum. Of these the Melian alone, and such others as possessed its qualities—that is, the opposite of those above ascribed to Samian earth—were used by painters.[3] There were in Melos and in Samos many varieties of earth. The Samian was obtained from a vein of considerable extent, but only two feet in height between the rocks which formed its roof and floor, so that one could not stand erect

---

[1] Dioscor., v., 181. It was also used by women as a cosmetic, and seems to have answered the same purpose as the cerussa, which is mentioned with it. (Plaut., Mostel., i., 3, 107.)

[2] Theoph., De Lapid., c. 108; Plin., H. N., xxxv., 19.

[3] Theoph., c. 107.

while digging it, but was obliged to lie upon his back or side. This vein contained four different qualities of earth, which became better in proportion as it was obtained from nearer the centre of the vein, the outer and inferior kind, called aster, being chiefly or solely used for cleansing garments.[1] Such of these white earths as were used for pigments resembled probably the clays still so employed under the names of Spanish white, white of Moudon or Morat, and Rouen white.[2] The white pigment, however, scraped from paintings in the tombs of the kings at Thebes appeared from Dr. Ure's analysis to be "nothing but a very pure chalk, containing hardly any alumina, and a mere trace of iron."[3] The Tymphaic earth was an earthy gypsum found near Perrhæbia and Tymphæa, in Ætolia, on the surface of the ground, and was used in cleansing garments.[4] It is remarkable that Pliny should have omitted all mention of

[1] Some suppose one variety of Samian earth to have been meerschaum, and that of this were formed those celebrated Samian vessels spoken of by Pliny (xxxv., 46), by Plautus (Capt. a, ii., sc. 2), and repeatedly by Cicero (ad Her., iv., 65; in Verr., i., 50).

[2] See Tingry's Painter's and Varnisher's Guide, p. 181, *seq.*

[3] Wilk., Anc. Egyp., vol. iii., p. 303.

[4] Theoph., De Lapid., c. 111; Plin., H. N., xxxvi., 59.

two earths which are spoken of by grave authors as applied to an important use. Varro speaks[1] of sprinkling "Chalcidicam aut Caricam cretam" among grain to preserve it, and Strabo, describing a public granary at Cyzicus, in Phrygia,[2] adds, "ποιεῖ δὲ τὸν σῖτον ἄσηπτον ἡ Χαλκιδικὴ μιγνυμένη." What these earths of Chalcis and of Caria may have been we must leave for others to determine.

OF MARBLES AND OTHER MINERAL SUBSTANCES EMPLOYED IN BUILDING, IN STATUARY, ETC., BY THE ANCIENTS.

In the earlier times especially, and generally in proportion as the difficulties of transportation were great, we find that the materials used in building were such as the neighborhood could most readily supply. The pyramids of Djizeh, the oldest existing structures, are of the limestone furnished by the opposite bank of the Nile, and which, during the inundation, could be deposited almost at their base. The temples of Thebes are of the sandstone which from no great distance above was floated down the Nile. The walls and private dwellings

---

[1] De Re Rust., i., 57, 1.

[2] Page 575.

throughout Egypt were of unburnt bricks, made from the earth at hand wherever they were wanted. In Greece, the Treasury of Atreus and the Citadel of Mycenæ, which for antiquity may almost rival with the earliest works of Egypt, are built of the calcareous conglomerate or pudding-stone quarried upon the spot.[1] The theatres of Chæronea and of Argos are formed against the hillsides by excavation of the living rock. The temples of Athens were of the marble of the neighboring Pentelicus. Some of the ornamental parts of the Erechthéum were of compact limestone, a material of which the use in Greece and elsewhere is more ancient than that of any marble.[2]

In Italy, besides that ready material, travertine, of which the temples at Pæstum are constructed, there abounded in the neighborhood of Rome various volcanic rocks, of which, as we shall see hereafter, the most ancient remains of that city are composed.

Among the marbles enumerated by Theophrastus and Pliny, that ranks first with both, which, from the island Paros, where it was ob-

---

[1] Dr. Clarke calls it a breccia, but no angular fragments are to be discerned.

[2] Clarke's Travels, vol. vi., p. 241.

tained, was called Parian, and from the manner in which it was quarried, by the light of lamps, was sometimes, as Pliny on the authority of Varro tells us, designated by the name lychnites.[1] This is the stone which, being regarded as best suited for statuary,[2] was used by Praxiteles and other illustrious Grecian sculptors.[3] Its whiteness is celebrated by Pindar[4] and Theocritus,[5] and its color, Plato thinks, is pleasing to the gods.[6] Of this marble are the Venus de' Medici, the Diana Venatrix, the colossal Minerva (called Pallas of Velletri), Ariadne (called Cleopatra), Juno (called Capitolina), and others. Of this are also the celebrated Oxford marbles known as the Parian Chronicle. Before Pliny's time, however, whiter marbles than the Parian had been discovered, and among them the Lunensian (marmor Lunense), which derived its name from the city Luna in Etruria, where it was obtained.[7] The same marble, now quarried at Carrara, a few

[1] From λῦχνος, a lamp.—Plin., H. N., xxxvi., 4, 2.
[2] Strabo, x., p. 711.
[3] Prop. iii., 9, 16; Quinct., ii., 19.
[4] Pind., Nem., iv., 131.
[5] Theocr., vi., 38; Clarke's Trav., vol. vi., p. 133.
[6] Plato, De Leg., xii., 7.
[7] Plin., H. N., xxxvi., 4, 2.

miles distant from its ancient source, is the one chiefly used by modern sculptors, and furnished the material for some fine specimens of ancient art likewise, as the Antinous of the Capitol and the Belvidere Apollo, according to Jameson, Dolomieu, and other mineralogists, though Dr. Clarke thinks them to be Parian marble.[1]

The Pentelican marble, which Theophrastus names next after the Parian, was obtained from Mount Pentelicus, not far from Athens, and many of the finest works executed there during the administration of Pericles are of this material. The Theséum and the Parthenon were built entirely of it, as was also the temple of Ceres at Eleusis;[2] but, being less homogeneous than the Parian marble, and consequently more liable to decomposition, the works executed in it do not, like those in Parian marble, retain the mild lustre of their original polish.

The Chian marble, which stands third in Theophrastus, was sometimes, as we may, from what Pliny says, infer, diversified with colored

---

[1] Trav., vol. vi., p. 136.

The author of "Rome in the Nineteenth Century" says the Belvidere Apollo "is now universally recognized to be of Italian marble."

[2] Clarke's Travels, vol. vi., p. 135, *seq.*

spots;[1] and from another passage in the same author, read correctly, it appears that the Lucullean marble also, which was of a uniform black, came from Chios;[2] and this pure black marble is indeed the kind most commonly spoken of as Chian, though it is not improbable that the island produced marble of several different sorts.

Beckmann thinks[3] that Theophrastus, by the carbuncle (ἀνθράκιον) of Chios,[4] meant this well-known black marble, which from its resemblance to an extinguished coal was designated by this name, as the ruby was from its resemblance to one burning; and of this marble he supposes to have been made the mirrors mentioned by Theophrastus, and that Pliny[5] misinterprets him in stating they were of the ἀνθράκιον of Orchomenus. The Thebaic, which is the only other marble that Theophrastus names, is not mentioned by Pliny, for his Thebaic stone and his Syenites of the Thebais were not marbles, as we shall hereafter see. Indeed,

---

[1] H. N., xxxvi., 5.

[2] The true reading appears to be "Chio Insula," instead of "Nili Insula." See Hardouin on Plin., H. N., xxxvi., 8.

[3] H. of I., iii., 178.

[4] De Lapid., c. 61.

[5] Plin., H. N., xxxvii., 25.

it is hard to say what the Theban marble of Theophrastus was, unless perhaps such a compact limestone as that in which the tombs at Biban-el-Molouk are excavated may, when polished, have been called marble. Champollion, however, found among the ruins at Sais fragments of white marble "dit de Thebes."[1]

Pliny deems it unnecessary to describe the kinds and colors of marbles, seeing they were so well known, and observes that it is not easy to enumerate them from their multitude.[2] We are now at a loss to determine even some of those he has thought fit to specify. We know that he includes among marbles some stones not strictly such, as porphyry, basalt, and syenite, and we are therefore justified in doubting as to others. Thus the Coraliticus, whose dimensions never exceeded two cubits,[3] would seem not to have been marble. If, however, it was, as some suppose, a snow-white fine-grained marble, which had its name from the River Coralus in Phrygia, Pliny must mean that it was not obtained from the quarry in masses of larger dimensions than those he has assigned to it. Corri[4] thinks it to be the marble now called

---

[1] Lettres d'Egypte, p. 52.
[2] H. N., xxxvi., 11.
[3] H. N., xxxvi., 13.
[4] Piet. Ant., p. 88.

palombino and colombino, of which are formed the cinerary vases numbered 1178, 1565 in the Galleria de' Candelabri of the Vatican. The Alabandicus, which "liquatur igni, ac funditur in usum vitri,"[1] is thought by some to have been marble used as a flux in making glass, while others have supposed it to be manganese.[2]

The Lacedæmonian green marble, which Pliny classes with the most precious, was quarried at Tænarus, and the verde antico is by some supposed to be the same.[3] Dr. Clarke says, "The Lacedæmonian was one variety of the verde antico, but it was green and black instead of green and white."[4] He finds the locality of the true verde antico in the defile of Tempe, on the side of Mount Ossa. He considers it the same with the marmor Atracium, so called from the city of Atracia, near which it was quarried. This is the opinion which Corsi also has adopted.[5] Bruce speaks of quarries of what he terms verde antico near Cosseir on the Red Sea.[6]

---

[1] Plin., H. N., xxxvi., 13.

[2] Beck., Hist. of Inv., iv., 60.

[3] Winck., Storia delle Arti del Dis., vol. i., p. 23; Sext. Empir. Op., p. 34, and the authors there cited by H. Stephens.

[4] Trav., vol. vii., p. 361.

[5] Piet. Ant., p. 94.

[6] Trav., vol. i., c. 8.

There are Egyptian green and white marbles still known, which are taken to be those Pliny calls Augustum and Tiberium. The ophites, with which Pliny compares them, is called in later writers Egyptian ophite, green Egyptian marble, and Egyptian green. It is by some regarded as a variety of common serpentine. Others describe it more accurately as a mixture of reddish-brown common serpentine, leek and pistachio green precious serpentine, white granular foliated limestone, and small portions of diallage.[1]

Of ophites there are three varieties specified by Dioscorides:[2] one black and heavy, a second ash-colored and spotted, the third containing white lines. The first was perhaps green porphyry, the ophites of Waller,[3] the second talc, and the third the kind just now described. Pliny seems to have in view the first two of these three varieties when he says there are two kinds of ophites, the one soft and white, the other dark-colored and hard.[4] He after-

---

[1] Jameson's Min., i., 154. This appears to be the stone described by Winckelmann under the name of Egyptian breccia. See Stor. delle Arti del Dis., vol. i., p. 89.

[2] Lib. v., c. 162. [3] Waller's Syst. Min., i., 430.

[4] H. N., xxxvi., 11.

ward speaks of a mortar for the physician's use made of the white ophite, and adds, "est enim hoc genus ophitis ex quo vasa etiam et cados faciunt;"[1] and so they do at this day in Germany, but more commonly of a dark-colored serpentine.

When luxury and the rage for building were at their height in Rome, every accessible region of the earth was ransacked for materials, and from the continent and islands of Greece, from Numidia and Egypt, from Phrygia and other parts of the coast of Asia Minor, the finest marbles were procured.[2] Among these were celebrated the Hymettian, from Mount Hymettus in Attica, the Carystian, from Eubœa, the white marbles of Cappadocia and Thasos, the black of Lesbos, and the marbles of Numidia, a country which, as Pliny asserts, produced nothing whatever remarkable except marble and wild beasts.[3] That quarried near Synnada in Phrygia, resembling alabaster in its variegated appearance, was much used at Rome in Strabo's time, and slabs and columns of extraordinary size and beauty were brought

[1] H. N., xxxvi., 43.

[2] See Hor., Od., ii., 18; Juven., Sat. xiv., 89; Stat., Theb., i., 144.

[3] H. N., v., 2.

from that great distance.[1] It is natural, therefore, that among the ruins of Roman architecture should be found marbles of which the quarries are no longer known, and which are therefore designated as antique. Such are the verde antico, the rosso antico, the giallo antico, the bianco e nero antico,[2] besides others.

The porphyrites quarried in the Thebaid[3] was no doubt the same stone elsewhere called pyropœcilus, pyrrhopœcilus, Syenites, lapis Thebaicus, etc. It derived the first name from its prevailing roseate tint, the next two from the fiery or the ruddy spots with which it was diversified, and was called Syenites and lapis Thebaicus from its being quarried near Syene, on the northern border of the Thebaid. The name *sienite* is now used to designate a rock which differs from granite in the substitution of hornblende for mica, but "it is plain that the Roman naturalist meant to designate by this term (Syenites) the common red granite found in abundance about Syene."[4]

[1] Strabo, p. 577.

[2] Dr. Clarke says the bianco e nero antico is an aggregate of dark diallage and white feldspar found in Macedonia.—Trav., vol. vii., pref., p. 18, and vol. viii., p. 15.

[3] Euseb., de Mart. Palæst., c. 8.

[4] Hamilton's Ægyptiaca, p. 8, note.

This material was profusely used in ancient Egypt in the construction or adornment of public edifices, in statuary, especially that of colossal size, and in forming those obelisks described by Pliny,[1] and still gazed at with wonder in Egypt or at Rome, in which city there remain at this day more of them perhaps than in all the world besides, Egypt, even, included.

This stone is rightly classed by Winckelmann with granite, of which he says Egypt furnished two varieties, one red and whitish, of which are formed these obelisks and many statues—as three of large size in the Capitoline museum—the other white and black, peculiar, as he thinks, to Egypt, and of which there is a statue of Isis in the same museum, and a large figure of Anubis in the Villa Albani.[2] Pliny speaks of fine works of art in Egyptian basalt also, and of these likewise some have found their way to Rome, as the lions at the base of the ascent to the Capitol and the sphinx of the Villa Borghese. Of this stone also Winckelmann distinguishes two kinds: the black, which is the more common sort, is the

[1] H. N., xxxvi., 13.

[2] Winck., Stor. delle A. del Dis., vol. i., p. 85; vol. ii., p. 13.

material of the figures just now mentioned; the other variety has a greenish hue.[1] This black basalt is the Æthiopic stone of Strabo, who speaks of it as used for making mortars. For statues of the Nile, which flows through Æthiopia, this was considered the most suitable material, though white marble was used to represent other rivers.[2] That famous statue of the Nile which Vespasian dedicated in the Temple of Peace was formed accordingly of this substance.[3]

The stone which Pliny describes as possessing the hardness of marble, white, and translucent, and thence called phengites,[4] is supposed to have been selenite.[5] Of this stone Nero built a temple of Fortuna Seia, within which, "when the doors were closed, there remained the light of day, which seemed not transmitted, as through the specular stone, but, as it were,

---

[1] Winck., Stor. delle A. del Dis., vol. i., p. 85; vol. ii., p. 13.

[2] Paus., Arcad., p. 647.

[3] Plin., H. N., v., 9. The sharp Æthiopic stone, which Herodotus (ii., 85) says was used in the process of embalming, to lay open the bodies of the dead, was probably flint, since many knives of that material *taken from the tombs of Thebes are preserved in European collections.* (Wilk., Anc. Egypt., iii., 262.)

[4] From φέγγος, splendor.

[5] Jam. Min., ii., 241.

inclosed within the walls."[1] Hardouin, in his note upon this passage, says the Church of St. Miniat in Florence is lighted in like manner by windows of selenite fifteen feet in height.[2] And it was thus, as Beckmann thinks, that selenite was used in the Temple of Fortune; the openings in the walls were closed, and not the whole building constructed with it.[3] Domitian, that he might be able to see what was done behind his back, lined with this same substance a gallery in which he used to walk.[4]

The tophus of Virgil[5] and Pliny[6] was a calcareous tufa, which, in some places a firm and durable material for building, was elsewhere, as

[1] Plin., H. N., xxxvi., 46.

[2] Hardouin has probably adopted this misstatement from Fontani, who says of this church, "Cinque finestre d'antica maniera comunicano meno sfacciata luce al presbiterio, essendo ornati con specchi di fengite, ossia pietra specolare." (T. i., p. 64.) The fact is that these five windows (if such they can be called) are filled not with selenite, but with a variegated substance looking more like marble than alabaster, of which the whiter portions are in some degree translucent, it is true, but transmit exceeding little light. The church receives what light it has from small glazed windows, placed at a considerable height.

[3] Hist. of Inv., iii., 175.

[4] Suet., in Vita Domit., c. 14.

[5] Georg., ii., 214.

[6] H. N., xxxvi., 48; xvii., 4.

at Carthage, of so spongy and friable a nature that it could not without some protection bear exposure to the weather, so that the Carthaginians, who had no other building material, used to coat their walls with pitch. A variety of calcareous tufa, which, though lighter than Parian marble, resembled it in hardness and in color, was called by the Greeks πῶρος or πώρινος λίθος, as it is by Pliny also when translating from Theophrastus.[1] Of this stone were constructed the temple of Jove at Elis,[2] and that of Apollo at Delphi,[3] except that the roof of the former was of Pentelican and the front of the latter of Parian marble. Indeed, porus seems to have been as usual a building material in Greece as the like substance, travertino (lapis Tiburtinus), was in Italy;[4] and these varieties of calcareous tufa were the first stones upon which the statuaries of Greece and Italy exercised their art.[5] The famous Belvidere Torso in the Vatican Museum, the work of Apollonius, is of this marmor porinum, a fine-grained variety of calcareous tufa. The stone

---

1 H. N., xxxvi., 28.

2 Paus., p. 397.

3 Herod., v., 62.

4 See Winck., Stor. delle Arti del Dis., vol. i., p. 22.

5 Ibid.

now commonly called *tufo* in Italy, and much used as a building material, is a volcanic production of a wholly different kind.[1] The lapis Tiburtinus, classed by Vitruvius[2] and Pliny,[3] as it generally is, with tufa, some have preferred to regard as a compact limestone.[4] It is sometimes so compact as to be well fitted for statuary, and susceptible of a fine polish—witness the salamanders on the front of the Church of S. Luigi de' Francesi in Rome. The travertino of Civita Vecchia, near the Terme Taurine, resembles Carrara marble. Of this plentiful material, travertino, are the ruins of Pæstum, and, in Rome, the Coliseum, St. Peter's, and numerous churches and other buildings, ancient as well as modern. In Strabo's time, indeed, the greater part of Rome was of πῶρος, or of travertino.[5]

It is a stony concretion, often of a spongy nature, and is deposited very abundantly from the waters of the Teverone. In the quarries of it fresh stone forms, in which tools are some-

[1] Corsi, Piet. Ant., p. 72.

[2] Vitruv., ii., 7.

[3] H. N., xxxvi., 48.

[4] That it was, however formed, a limestone, we should know from Palladius, who speaks of it as capable of being burned to lime.

[5] Strabo, v., 238.

times found imbedded.[1] There is an example of a like extensive formation of tufa near Guanca Velica in Peru, where a warm spring deposits calcareous earth in such abundance as to constitute quarries, from which building materials are drawn.[2] Limestone of this kind acquires a great degree of hardness, and it has been thought that the solidity of Roman masonry is due to the joint use of travertino and puzzolana, which are found in the same neighborhood.[3]

That puzzolana, mixed with a small proportion of lime, hardens quickly, even under water, was well known to the ancients. Vitruvius, Pliny, Seneca, and others speak of this property of the pulvis Puteolanus as a curious fact.[4] The black sand which Vitruvius calls carbunculus, and supposes to have been produced by subterranean heat, was probably the black variety of puzzolana, which some mineralogists regard as altered scoria.

The existence of such subterranean heat Vitruvius thinks is evidenced in various ways, as

---

[1] Kidd's Min., i., 25.

[2] Brongn., T. E. de Min., i., 212.

[3] Ibid.

[4] See Vitruv., ii., 6; Plin., xxxv., 47; Senec., Nat. Quæst., iii., 20.

by warm springs, by hot vapors issuing from hollow mountains, by ancient traditions of the effects of such heat—effects which, he thinks, must vary with the nature of the soil; wherefore the same heat produced in Campania the dust (pulvis) called Puteolanus, and in Etruria that sand called carbunculus.[1] The pulvis Puteolanus was obtained from the region round about Vesuvius, and especially near Baiæ and Puteoli, now called Puzzuoli, whence its modern name.

That stone into which earth from the neighborhood of Cyzicus and Cassandria, when immersed in the sea, was changed, as Pliny tells us,[2] as also that along the coast from Oropus to Aulis, where the earth as far as it was washed by the waves was converted into rock, may have been a sort of calcareous tufa, or a calcareous breccia, like that on the coast of Guadaloupe, in which are imbedded human bones.[3] There are other instances of similar aggrega-

---

[1] Vitruv., ii., 6, sub fin. It should be remembered that Vitruvius wrote before the first eruption of Vesuvius of which we have any historic record—that by which Herculaneum and Pompeii were destroyed.

[2] Plin., H. N., xxxv., 47.

[3] Lyell's Princ. of Geol., vol. iii., p. 336.

tions from the same cause. "The delta of the Rhone has been almost wholly converted into a solid rock by a calcareous cement."[1] The spring in Cnidus which in eight months hardened earth into stone was probably impregnated with carbonate of lime to such a degree as to produce effects like those exhibited at Matlock in England, at Carlsbad in Bohemia, at the hot baths of St. Philip in Tuscany, or, in ancient times, near Hierapolis in Phrygia, where, as Vitruvius informs us,[2] was a copious hot spring, the waters of which, introduced into trenches cut around gardens and vineyards, deposited a crust of stone—a property of which the inhabitants who dwelt thereabout took advantage in order to form inclosures for their fields.

Strabo, too, speaks of these waters of Hierapolis, "which were so easily indurated into calcareous tufa (πῶρος) that, by conducting them through channels, hedges were formed of one entire stone;"[3] and those who have noticed the rapid progress and extent of similar formations at Tivoli and elsewhere in Italy will easily believe this possible.

The silices, of which certain kinds are speci-

---

[1] Renwick's Outlines of Geology, p. 34.

[2] Vitruv., viii., 3, 10.

[3] Strab., p. 629.

fied by Pliny[1] as fit to be used in building, may in some cases have been substances which we also term silicious; but the more probable opinion is that the name silices was somewhat indiscriminately applied to the more compact and harder stones.

The viridis silex, which so remarkably resisted fire, which was never abundant—a stone rather than a rock—may perhaps have been serpentine. No inference to the contrary need be drawn from Pliny's calling it silex, for he presently after speaks of lime made "ex silice," as Vitruvius also directs that it be burned "de albo saxo aut silice." It is probable that by silex in these passages is meant a dark-colored compact limestone.

The red stone which Pliny,[2] after Vitruvius,[3] recommends to expose to the weather for two years before it is used in building, in order that such stones only may be employed in the superstructure as have been able to abide this test, may have been a red sandstone, different varieties of which differ greatly as to their durability when so exposed.[4] But it is more

[1] Plin., H. N., xxxvi., 49. [2] Id., ib., xxxvi., 50.

[3] Vitruv., ii., 7.

[4] Some, as that from Nyack on the Hudson, absorb moist-

probable that they mean the stone procured from quarries at a place called Saxa Rubra by the ancients, now Grotta Rossa.[1] Some beds of this stone have been explored to a depth of fifty feet, and new discoveries are constantly made of ancient excavations. This red stone is a sort of indurated puzzolana, and consists, like the Alban stone, of volcanic cinders which have been diffused in water and deposited in strata.[2] It varies in color, but is commonly of a dull pale red. Of this reddish tufa are the Tarpeian rock, the lower part of the steps that lead up to the temple of Vesta, the columns of the temple of Hercules Custos within the convent of S. Niccola a' Cesarini, the outer walls and many of the middle columns of the temple of Fortuna Virilis. The lapis Albanus, which Vitruvius is the first to mention, was so called from its origin, the volcanic craters of the Alban mountains. It is composed of cinders and pebbles that have become indurated into a solid

ure and fall to pieces, while others, as that from Newark on the Passaic, become harder by exposure to the weather.

[1] See Livy, ii., 49; Tacit., Hist., iv., 79; Cic., Philip. ii., 31; Aur. Vict. de Cæs., c. xl., 23; Strabo, lib. v. This last-named author says the Anio flows near the quarries of Tiburtine and Gabian stones, and that called Red.

[2] Corsi, Piet. Ant., p. 71.

mass. Its Italian name, peperino, is derived from the resemblance which some of the various matters disseminated through it bear to grains of pepper. No ancient building of this material has withstood the effects of time, but there are statues, basins of fountains, vases, and various other things formed of it, that still remain in a good state of preservation. The much-admired sarcophagus of Luc. Corn. Scipio Barbatus, in the first chamber of the Museo Pio Clementino, is of this Alban stone.[1] The lapis Gabinus, now called sperone, derived its name from its quarries being near the Lake of Gabinus, now called Lago di Castiglione, the crater probably of the volcano to which it owed its origin. It is commended by Tacitus as well resisting fire.[2] It is harder than the Alban stone, as containing a greater proportion of pebbles and silicious fragments than of cinders. Of this material are formed the arch of the Cloaca Maxima, the Tabularium of the Capitol, and the Arco de' Pantani, as it is called, a part of the Forum of Nerva.[3] A still harder volcanic stone was the lapis Anitianus, mentioned by Vitruvius,[4] if, at least, that was the

[1] Corsi, p. 67.
[2] Ann., xv., 43, Præf., i., Hist.
[3] Corsi, p. 70.
[4] Vitruv., i., 7.

granitic lava now called manziana — a corruption, as Corsi supposes, of its ancient name.[1]

The silex Tusculanus spoken of by Pliny,[2] and alluded to by Vitruvius,[3] was probably the basaltic lava formed by the volcano of which Lake Regillus, comprehended within the Tusculan territory, is regarded as the crater. With this stone were paved the Viæ Sacra, Ostiensis, Aurelia, Appia, and others.[4]

Pliny specifies various kinds of sand used in sawing marble, of which the best were the Æthiopic and the Indian. Of sand used in making mortar he afterward distinguishes three sorts—fossil, river, and sea sand (fossitia, fluviatilis, and marina); and in relation to the former he has a brief geological remark, which seems borrowed from Vitruvius, who says that fossil sand is not wanting in the parts of Italy which lie west of the Apennine Mountains; but that beyond them, toward the Adriatic Sea, there is none found, and that in Achaia, in Asia, and, in fine, throughout the countries beyond sea, it is not even named. This assertion, however, Philander, Scamozzi,

[1] Piet. Ant., p. 73.
[2] H. N., xxxvi., 18.
[3] Vitruv., i., 7.
[4] Corsi, p. 74.

and other commentators on Vitruvius say is incorrect.[1]

OF SALINE SUBSTANCES MENTIONED BY THE ANCIENTS.

Notwithstanding Beckmann's learned and labored argument to prove the ancients unacquainted with our alum, and that the salt called by them alumen was a native sulphate of iron,[2] there are some who still think that the manufacture of alum must have been known at a much earlier period than he assigns for its first introduction,[3] or else that it was found native in Egypt, Melos, Lipari, and elsewhere, in sufficient quantity for the purposes to which it was anciently applied.

We know that the ancients, without much discrimination, often called by the same name substances widely different in their nature, but which in certain points bore some resemblance to each other; and it is probable the *στυπτηρία* of the Greeks, called alumen by the Romans, may in some instances have been sulphate of iron, or some other vitriolic salt besides what

---

[1] Vitruv., ii., 6, and annotations thereon.

[2] Beck., Hist. of Inv., i., 288.

[3] See Parke's Chem. Essays, vol. i., p. 625.

we call alum; but that this too, or a salt very nearly resembling it, was sometimes meant seems certain.

Some of the properties ascribed to alumen belong to our alum, but not in the least to sulphate of iron. Thus Dioscorides says[1] of the fibrous variety, which he terms σχιστὴ, that it is exceeding white (λευκὴ ἄγαν), and from Pliny's description of this kind[2] we may infer that it was such an efflorescence as that called plume, or fibrous alum, found in considerable quantity in cavities and fissures of the earth at Solfatara near Naples, and at La Tolfa, in alum-stone.[3] Pliny, too, speaks of alum as white—as used for whitening wool—as being prepared for certain medicinal uses by burning it in pans until it cease to flow liquid—all which is quite inapplicable to sulphate of iron.[4] The modern alum of commerce generally contains more or less sulphate of iron,[5] and the alumen of the ancients, whether native or prepared according to their unskillful methods, may have contained so much of the same impurity occasionally as

[1] Dioscor., v., 123. [2] H. N., xxxv., 52.

[3] Cleav. Min., p. 227.

[4] Vide Mill. in Celsum, lib. iv., c. 18.

[5] Cleav. Min., p. 229.

to blacken when mixed with the juice of the pomegranate or with galls.[1] Besides, they may have confounded with alumen what by some of the earlier mineralogists of modern times was called hair-salt, and supposed to be a variety of alum, but which analysis has shown to be a mixture of the sulphates of magnesia and iron.[2]

As this is called by Kirwan hair-salt,[3] so Dioscorides applies to a certain variety of στυπτηρία a corresponding name, τριχῖτις.[4] Of the many varieties which he says there were, he has specified three as used in medicine: the fibrous or separable (σχιστὴ), the round (στρογγύλη), and the moist (ὑγρά). The σχιστὴ may be rightly denominated *fibrous*, since he says it greatly resembled a certain stone, which might be distinguished from it, however, as possessing no astringent taste; and this stone, we afterward discover, was amianthus.[5] This comparison, altogether unsuitable if we suppose στυπτηρία σχιστὴ to be sulphate of iron, is perfectly just as applied to what has been in modern times called fibrous alum, which is "composed of capillary crystals or fibres, white, silky, parallel to

[1] Plin., H. N., xxxv., 52.
[2] Cleav. Min., p. 227.
[3] Kirwan's Min., ii., 13.
[4] Dioscor., v., 123.
[5] Dioscor., v., 156.

each other,"[1] while "the fibres of amianthus are usually straight, almost always parallel, have in most cases a silky lustre, and their color white."[2]

In the island Milo are found, mingled with fibrous gypsum in volcanic rocks, fine specimens of what till lately was thought fibrous alum,[3] but which appears upon analysis to be a sulphate of alumina containing accidental portions of potash, soda-alum, sulphates of magnesia, iron, and copper.[4] This island is mentioned very particularly by Dioscorides as a locality of *στυπτηρία*, and Pliny observes that the best alumen of a certain kind was that called Melinum, from the island Melos,[5] having before said of the substance generally that, after the Egyptian, the most approved was found in Melos. There can be little doubt, therefore, that the alumen of the ancients comprehended at least this sulphate of alumina, or rather that this is the substance generally meant by the

[1] Cleav. Min., p. 227. [2] Ib., p. 405. [3] Ib., p. 227.

[4] See Amer. Jour. of Science and Art, No. xlvi., p. 388. Klaproth, however, found the requisite proportion of potash in the alum of the Grotta di Alume, at Cape Miseno, near Naples, "where nature alone," he says, "unassisted by art, is constantly producing perfect alum." (Min., i., 266.)

[5] Plin., xxxv., 52.

words στυπτηρία and alumen; for the islands of Lipari and Milo, where the sulphate of alumina is still produced in considerable abundance, were in the times of Diodorus Siculus and Pliny the chief sources of alumen.[1]

The alum mines of Milo, one of which Tournefort examined, would appear from his description[2] to have been capable of furnishing no inconsiderable quantity; but when Diodorus wrote, it seems the supply from this quarter, whatever it may have been, was of small account in comparison with that derived from Lipari, which island, he remarks, contained the celebrated mines of alum (στυπτηρίας) whence the Lipareans and the Romans drew large revenues; for, enjoying a monopoly of a substance much used, and nowhere else produced, except in small quantity in Melos only, and raising the price at pleasure, their profits were to an amount incredible.[3] Of Egypt, mentioned by Pliny as a locality of alumen, Diodorus takes no notice, but Herodotus relates that Amasis sent to the people of Delphi, as his contribu-

[1] See Penny Mag., vol. iii., p. 191, respecting the abundance of native alum in Milo.

[2] Letter iv., vol. i., p. 128, 129 of Eng. trans.

[3] Diod. Sic., v., 10.

tion toward the building of their temple, a thousand talents of this mineral,[1] which may have been derived from the neighborhood of Syene, where native alum is at this day found.[2]

The kind which Dioscorides and Pliny call στρογγύλη, or round, may have been such concretions or small fibrous masses of native alum as are occasionally found, or rather, perhaps, aluminite, or subsulphate of alumina, which occurs in roundish or reniform masses, and, possessing some properties in common with alum, would, like hair-salt, have been readily confounded with it. Some of the characters of that called ὑγρὰ, or moist, by Dioscorides agree well enough with those of stone or mountain butter, as Kirwan, Werner, Jameson, and others term a certain variety of native alum. If we must, with Pliny,[3] call it liquid, it may have been a strong solution of the native salt, found occasionally in cavities formed in aluminous slate, or in clays embracing the sulphuret of iron.

By *chalcitis*, "a stone from which copper was

[1] Herod., ii., 180.
[2] Brard, Min. appliquée aux Arts, vol. i., p. 309.
[3] H. N., xxxv., 52.

melted,"[1] Pliny no doubt means pyritous copper; but the relation this bears to sulphuret of iron (such that it is sometimes hard to distinguish between the two)[2] caused them to be confounded, and consequently the sulphate of iron procured from and found associated with either of these minerals was called χαλκάνθον, chalcanthum, flos æris, copperas; and as the two ores were easily confounded, we need not wonder at a like confusion as regards the two sulphates that are derived from them; and again, since the sulphate was sometimes found associated with the sulphuret of iron, and that was not distinguished from pyritous copper, we find the characters and properties of all the three confounded in one and the same description. Thus Pliny describes[3] chalcitis as a stone from which copper is extracted—meaning pyritous copper; as soft, friable, and like a downy concretion—a description applicable to the sulphuret of iron, as it occurs at times in efflorescences, or stalactitical concretions, or in crusts composed of fibres. The three kinds into which he says chalcitis is distinguished, viz., brass, and misy, and sory, are again either py-

[1] Plin., H. N., xxxiv., 29. [2] Cleav. Min., p. 558.
[3] H. N., xxxiv., 29.

ritous copper or the sulphuret of iron, and the sulphate of iron in two different states. When he says it has veins of brass, and is most approved when of the color of honey, and with delicate branching veins, friable, not stony, he seems to describe that sulphuret of iron which occurs in membranes or dendritic branches embraced between layers of other minerals.

The styptic and caustic properties which Dioscorides assigns to chalcitis belonged of course to the vitriolic salts derived from it.[1] The sulphate of iron becomes from exposure yellow, and at last brown, and in these states probably answers to the misy and sory of Dioscorides and Pliny.[2] Some of the characters which they ascribe to these substances might, indeed, lead to a belief that they were varieties of pyrites, did we not know how apt these writers were to fall into the confusion of ideas and terms that has just been mentioned. That misy was what has been called Roman vitriol, or yellow copperas, seems unquestionable,[3] and it is expressly so named by Democritus,[4] who

---

[1] Dioscor., v., 115.

[2] Kidd's Min., ii., 22; Jam. Min., ii., 343.

[3] See Hardouin on Pliny, vol. ix., p. 320.

[4] As cited by Saumaise—Exercit. Plin., p. 815, a AB.

says that sory (σόρυ) is always found in misy (μίσυ), and that this latter is called yellow copperas (χλωρὸς χαλκάνθος).

Of the *melanteria*, or ink-stone, Dioscorides remarks that "some have taken it to be the same with sory, from which it is distinct, though not unlike."[1] It was probably a sulphate of iron formed in a matrix containing vegetable astringent matter, by union with which was produced a natural ink.

Chalcanthum, which appears to have been so named for the reasons above stated, rather than, as some suppose, because of the brass-yellow color of the mineral from which it was obtained, was otherwise called atramentum sutorium, from its being used to blacken leather.

Pliny, describing a process by which it was made, says that upon cords suspended in the vats containing water impregnated with it there accumulated vitreous clusters of a remarkable lustre, and adds, "vitrumque esse creditur."[2] Hence are derived our terms vitriol and vitriolic.

That the ἅλς ἀμμωνιακὸς, sal ammoniac of Dioscorides, Celsus, and Pliny,[3] was not our

[1] Dioscor., v., 119. [2] H. N., xxxiv., 32.

[3] See Dioscor., v., 126; Cels., 66; Plin., H. N., xxxi., 39.

sal ammoniac, the chloride of ammonium, but rock salt, seems to have been sufficiently proved by Beckmann;[1] but his learned examination of this subject has not rendered it equally certain that the ancients were unacquainted with the chloride of ammonium. The rock salt designated as ammoniac was brought through Egypt, from as far, says Pliny, as the oracle of Hammon, where Arrian states that it was dug, a native fossil salt (*ἅλες αὐτομάτοι ὀρυκτοί*);[2] and when afterward the chloride of ammonium came to be manufactured in Egypt this received the same name. Athenæus speaks of sal ammoniac and water of the Nile sent from Egypt to the King of Persia, as though they were both alike intended for the use of the table.[3] The Tragesæan salt, mentioned by the same author,[4] appears to have been obtained from salt springs in the Troad, which Strabo calls *Τραγεσαῖον ἁλοπήγιον*.[5] It has been thought, however, by some modern mineralogists that the ancients knew chloride of ammo-

---

[1] Hist. of Inv., iv., 360, *seq.*

[2] De Exped. Alex., lib. iii., c. 4.

[3] Athen., i., 257.

[4] Vol. i., p. 288.

[5] See Poll. Onomas., vi., 63; Plin., H. N., xxxi., 7, 41, and the remarks there of Hardouin.

nium under the name of nitrum;[1] and although Beckmann maintains the opposite opinion, the grounds on which he rests his argument do not always bear him out. He observes that "there are two properties with which the ancients might have accidentally become acquainted, and which in that case would have been sufficient to make known or define to us this salt (the chloride of ammonium). In the first place, by an accidental mixture of quick-lime, the strong smell or unpleasant vapor diffused by the volatile alkali separated from the acid might have been observed." Now what Beckmann seems willing to admit as a criterion of sal ammoniac is mentioned by Pliny of nitrum, which, he says, sprinkled with lime, gives forth a powerful odor (calce aspersum reddit odorem vehementem).[2]

Beckmann appears to doubt what he says "several writers have asserted, that sal ammoniac comes also from the East Indies;"[3] but it certainly is brought thence at this day, and may have been manufactured there and have found its way thence to Europe in the time of Pliny also — for we find that unchangeable

[1] See before, p. 21. [2] Plin., H. N., xxxi., 46.
[3] Hist. of Inv., iv., 381.

country producing the same things then as now: indigo, Indian ink, fine steel, sugar, silks, etc.[1]—since "in India all artificial productions are of very great antiquity."[2] The manufacture of sal ammoniac in Egypt also may, for aught we know, have been more ancient than is thought. We are not justified in concluding that the ancients were ignorant of every thing of which we discover no mention in their works. Beckmann, on no other ground, rejects the sense put upon the words "usus ad fenestras" by Lehmann (who supposes Pliny here to speak of window-frames, to which a certain pigment was applied), and asserts that "glass windows were at that time unknown."[3] But now we find in Pompeii glass windows,[4] which we are very sure date as far back as Pliny's time, since he witnessed the catastrophe that buried them. One of the chief reasons for supposing the ancients to have been ignorant of our sal ammoniac and nitre is that we know of few uses to which they might have been ap-

[1] See before, p. 58, 89; Plin., H. N., xxxii., 17; xxxv., 25; xxxiv., 41; Virg., Georg., ii., 121.

[2] Beck., Hist. of Inv., iv., 118.

[3] Ibid., ii., 237.

[4] Lib. of Entert. Knowl., Pompeii, i., 155; Œuv. de Chateaubr., t. xiv., p. 163; Lettre de M. Taylor à M. C. Nodier.

plied. But though they may have had little inducement to manufacture them, even had they possessed the art, yet they could hardly have failed to observe them in a native state, since both these salts are found occurring thus in southern Italy and elsewhere.[1] Beckmann observes that the ancients were so much inclined to look for medicinal virtue in all natural bodies, there is reason to think they soon collected and made trial of the nitrous efflorescence sometimes found on walls.[2] Is it probable, then, they would neglect the purer nitre found in such caves as that at Molfetta in the kingdom of Naples? where, as appears from Klaproth's account of it, nitre is produced in great quantity,[3] as sal ammoniac is at Solfatara.[4] These two salts, therefore, the chloride of ammonium and the nitrate of potash, may be included among the varieties of nitrum spoken of by Pliny, although the greater part of them no doubt contained in some shape or other the carbonate of soda, because for the many and important uses to which that salt

---

[1] Cleav. Min., p. 122–124, 125.
[2] Hist. of Inv., iv., 529.
[3] Klapr. Min., i., 270; Cleav. Min., p. 124.
[4] Brongn., Tr. Elem. de Min., i., 110.

was anciently applied, great quantities of it must have been required.

The use of it in washing has been before referred to,[1] and Pliny mentions a great variety of purposes in medicine and in the arts for which it was employed.[2] In the manufacture of glass, according to him, three parts of nitrum were used with one of sand. A prodigious quantity of it must have been consumed in Egypt, since, according to all three modes of embalming described by Herodotus, the body was covered in this substance for the space of seventy days.[3]

Athenæus mentions *νίτρον* among various condiments, such as cheese, thyme, cumin, etc.,[4] but as drinking-vessels were cleansed with it,[5] that may have been the purpose to which it was destined when found among the supplies of the table.

Pliny's account of nitrum may supply an instance of the manner in which the ancients frequently distinguished between things essentially the same, because of some accidental dif-

---

[1] See *ante*, p. 21. [2] H. N., xxxi., 46.

[3] Herod., lib. ii., c. 86–88. Nitrum was obtained from the natron lakes in great abundance.

[4] Athen., i., 262; ii., 77. [5] Ib., xi., 108.

ference, and comprehended under one name things essentially different, because they happened to resemble each other in something accidental. We can not, therefore, observe strictly Beckmann's rule, "that those who wish to explain the old names of natural objects must relate every thing said of them, and not that alone which is favorable to their opinion,"[1] since we find ancient authors predicating of a name qualities that can not be combined in any one natural object; and Beckmann himself has before observed[2] that, "not being acquainted with any accurate method of separating and distinguishing salts, it need excite no wonder that they should ascribe to their nitrum properties which could not possibly be united in a salt."

The *melitites lapis* of Pliny[3] is supposed by some to be borax, the biborate of soda,[4] which may serve as our excuse for considering it and the minerals related to it under this head of saline substances. Dioscorides describes it as precisely like galactites, except that it emitted a sweeter juice;[5] and of galactites he says that it was of an ashen color, sweet taste, and yield-

[1] Hist. of Inv., iv., 534.
[2] Ibid., iv., 531.
[3] H. N., xxxv., 33.
[4] Kidd's Min., ii., 10.
[5] Dioscor., v., 151.

ed a milky juice.[1] This last the commentators understand to mean simply that its trace was white,[2] though the language of the description and the names of these related minerals would seem to indicate something more.

Another substance closely related to melitites and galactites was morochthus, by some called galaxias, and by others λευκόγραφις. This was found in Egypt, soft, easily soluble, and used in whitening linen.[3] It may be gathered from Dioscorides and Pliny, with the authors cited in the notes of Hardouin, that galaxias, galactites, morochthus, moroxus, morochites, leucogæa, leucographia, leucographis, and synophites differed in little except name, or were in fact varieties of the same substance, which came either from the Nile or the Achelous; was ash-colored, or greenish, or leek-colored, sometimes with red and white veins; was readily soluble, and when rubbed on stone or a rough garment left a white mark; besides which, when dissolved, or when triturated in

---

[1] Dioscor., v., 150.

[2] Vide Jani. Anton. Sarac. scholia in Dioscor., v., 152, et Hardouin annot. in Plin., H. N., xxxvii., 63.

[3] See Dioscor., v., 152, and Hardouin on Plin., xxxvii., c. 59, 63.

water, it appears to have resembled milk in color and in taste. The thyitis-stone of Dioscorides appears to be still much the same thing under a different name. It was found in Æthiopia, was greenish, resembling jasper, and when dissolved had a milky look.[1]

Now, minerals that answer the above description tolerably well are Spanish chalk and certain other varieties of talc, which are found of the colors indicated, may be mixed with and held suspended in water, so as to give it a milky appearance and a smooth, sweetish taste, and which, moreover, make a white mark when rubbed upon stone or cloth. It is from this property that those names leucogæa, leucographia, and leucographis are derived; and, supposing the mineral in question to have been talc, another of its names, synophites, that is, the associate of ophites, would be very naturally accounted for—the ancient ophites comprehending three different minerals, one of which appears to have been magnesian, and closely related to what is now called serpentine.[2] It is possible, however, that the substances we are considering were no more of mineral nature than were those which Strabo speaks of as

[1] Dioscor., v., 154. [2] See before, p. 108.

among the wonderful productions of India—"stones dug in that country, of the color of frankincense, sweeter than figs or honey;"[1] for these supposed stones were, in all probability, nothing else than sugar, of the nature of which Strabo, or those from whom he adopts his description, may easily have been ignorant, though Pliny was so far acquainted with *saccharon* that he describes it as a honey procured from reeds, white, resembling gum, friable under the teeth, never larger than a filbert, and used only in medicine.[2]

OF COMBUSTIBLE MINERALS MENTIONED BY THE ANCIENTS.

Θεῖον, *Sulphur*, was applied by the ancients to various uses in medicine and other arts. For the use of the physician was required translucent native sulphur, which the Greeks termed ἄπυρον.[3] That which had been freed from impurities by an artificial process—which had passed the fire—was called πεπυρωμένον, and distinguished into various kinds, appropriated to various uses, according probably to their

[1] Strabo, p. 703.

[2] H. N., xii., 17. He describes the sugar-candy still commonly used in India.

[3] Dioscor., v., 124.

several degrees of purity. Thus one kind was used for fumigating woolens, to render them more white and soft; another for making matches—purposes to which sulphur yet continues to be applied. The use of it in expiation and lustration, which was very common, we find referred to by many ancient authors.[1]

Ἄσφαλτος, *Bitumen*, in various states, from that of fluid, transparent naphtha to that of dry, solid, black asphaltum, was well known and much used among the ancients. Semiramis, we are told, employed it as a cement in building the walls of Babylon,[2] for which purpose it was obtained from the River Is, which discharged itself into the Euphrates about eight days' journey from Babylon.[3] But it was also found near that city in great abundance, and in a liquid form,[4] while elsewhere, as in Syria, asphaltum was dug from quarries in a solid state.[5] Its medicinal virtues may be found de-

[1] See Hom., Od., χ., 481; Theocr., Id. xxiv., v. 94; Ovid, A. A., lib. ii., v. 330; Juven., Sat. ii., v. 157.

[2] Diod. Sic., ii., 7; Vitruv., viii., 3, 8.

[3] Herod., i., 179.

[4] Strab., p. 743; Plutarch., vol. i., p. 685, D; Diod. Sic., ii., 12; Justin., i., 2; Quint. Curt., v., 1; but see, for a learned examination of this subject, Parkinson's Organ. Rem., vol. i., p. 131.

[5] Vitruv., viii., 3, 8.

tailed by Dioscorides[1] and by Pliny,[2] who also specify various economical uses to which it was applied. At Agrigentum, they inform us, it was burned in lamps as a substitute for oil, and called, in fact, Sicilian oil, but erroneously, as Dioscorides observes, it being a species of liquid bitumen (ἀσφάλτου ὑγρᾶς εἶδος).

A yellowish naphtha, found near Amiano in the Dutchy of Parma, has been in like manner used to light the streets of that city and of Genoa, and in Persia, also, a like substance is burned in the place of oil in lamps. A variety less fluid and transparent, called petroleum, is applied to the same use in different parts of the United States, as near Scottsville in Kentucky, where it was collected from the surface of a spring of water in sufficient quantity to be sold at twenty-five cents a gallon.[3]

The Egyptians used much asphaltum in the embalming of their dead.

The exceeding inflammable nature of naphtha, and even of its vapor, caused it to be employed by ancient jugglers in their deceptions with fire; and Plutarch[4] and Pliny[5] explain

---

[1] Lib. i., c. 99, 100, 101. [2] H. N., xxxv., 51.

[3] Cleav. Min., p. 488. See also Haüy, Min. iii., 224.

[4] Vol. i., p. 686, A. [5] H. N., ii., 109.

the story of the bridal dress and crown which Medea presented to Creüsa, by supposing them to have been anointed with naphtha, which, as soon as the princess approached the burning altar, burst out into flame.

The principal ingredient in the celebrated Grecian fire is supposed by Klaproth to have been some variety of asphaltum.

Ἤλεκτρον, *Succinum*, *Amber*, was well known to the ancients many centuries before the age of Pliny, and various ornamental articles were made of it, but in his time only for the use of women.[1] In Homer's verse,[2]

"Χρύσεον ὅρμον ἔχων μετὰ δ' ἠλέκτροισιν ἔερτο,"

it is probable that amber beads or ornaments at the ends of the necklace are meant by the word ἠλέκτροισι. The use of the plural number and the purpose to which the things spoken of were applied lead us to suppose this rather than that they were of the alloy called ἤλεκτρον.

Amber, according to the common fable, consisted of the tears of those poplars into which Phaeton's sisters were transformed;[3] but Pliny

---

[1] H. N., xxxvii., 11. [2] Od., v., 460.

[3] Dion. Perieg., v. 289, 316.

has collected a great variety of opinions of different authors as to its origin, its nature, and the derivation of its name, each one more extravagant than that which had preceded. His own belief, not differing much from the now received opinion, is that it consists of the resinous juice of certain trees, which had in course of time become mineralized in the earth. Hence was its Latin name succinum derived, "quod arboris succum prisci nostri credidere."[1] Its Greek name, *ἤλεκτρον*, was from its resemblance in color to an alloy of gold and silver so called, or from a name of the sun, Ἠλέκτωρ. The women of Syria, who ornamented their spindles with it, called it *ἅρπαξ*, because of its attracting leaves, chaff, and the fringes of their garments. Pliny says the different colors it exhibited in its native state were sometimes produced by artificial means, since they could dye it of whatever tint they pleased; wherefore it was much used in counterfeiting translucent gems, and especially the amethyst. Demostratus[2] called amber lyncurion, supposing it produced from the urine of the lynx—from that of the males when of a deeper and more

[1] H. N., xxxvii., 11.

[2] As cited by Plin., H. N., xxxvii., 11.

fiery tint, but when feebler and paler of the other sex. Other writers spoke of lyncurion as a substance distinct from amber, but having the origin indicated by its name. As Pliny, however, doubts[1] the existence of any such substance, we might perhaps spare ourselves the trouble of considering what may have been the mineral to which so strange an origin has been assigned. From what Theophrastus says of it,[2] taken in connection with the remarks of Pliny, it seems highly probable that some variety of amber had been distinguished by the name λυγκούριον. Theophrastus, it is true, compares lyncurion with amber, as if it were something different, though like; but they may, notwithstanding, have been one and the same thing, for he evidently speaks on the report of others, introducing his mention of it with the phrase "they say" (οἱ δὲ φάσι). It was engraved, he says, for seals, and that alone may have led him to ascribe to it the property of hardness, which would seem inconsistent with its being amber. He observes that, like amber, it attracts not only chaff and wood, but, as Diocles asserts, even light particles of brass and iron. It has been thought belemnite by

[1] H. N., xxxvii., 13.

[2] De Lapid., c. 50.

some, tourmaline by others, and by others again the hyacinth.[1] The ligure, in the description of the breast-plate (Exod., xxviii., 19), was probably this stone.

That distinction of sex, according to which Theophrastus, Pliny, and other ancients termed varieties of the same mineral male or female, as their color was more or less intense, or as they possessed in greater or less degree some other distinguishing property, is rather a fanciful one when applied to the sard, the cyanus, and others,[2] but as regards the lyncurion seems to have very naturally originated from the difference of sex in the animals supposed to have produced its different varieties.

There can be little doubt that the gagates of Dioscorides[3] and Pliny[4] is the modern jet, which with some mineralogists still retains its ancient name—a name derived from the River Gagas, in Lycia, about whose mouth this mineral was found.

The Thracian stone, which Nicander[5] directs to be burned together with gagates, sulphur,

---

[1] Beck., Hist. of Inv., vol. i., p. 141, 142.

[2] Theoph., c. 56. Miner. des Gens du Monde, p. 248.

[3] Lib. v., c. 146.

[4] H. N., xxxvi., 34.

[5] Theriac., v., 45.

bitumen, and other substances which during their combustion give forth a powerful scent, in order to drive away serpents, was brought, as that author and Dioscorides[1] inform us, from the Thracian river Pontus. Dioscorides ascribes the same virtues to it as to the gagates, and he and Pliny[2] observe of it, as Nicander does, that its flame is brightened by water but extinguished by oil,[3] though the two first named omit to add, when it is burning feebly—a condition which may be inferred from Nicander's words, "*τυτθὸν ὅτ' ὀδμήσηται.*" This property of being kindled by water and extinguished by oil, assigned by Dioscorides to asphaltum also, and by Pliny to gagates, might with about equal truth be ascribed to our bituminous coal, with which some naturalists have been disposed to identify this Thracian stone;[4] and, from the manner in which Aristotle speaks of it,[5] they seem to have good reason for so doing.

Theophrastus, speaking of combustible minerals, says that "some of the more frangible are

---

[1] Lib. v., c. 147.
[2] H. N., xxxiii., 30.
[3] Theriac., v. 46:

> "Ἡ θ' ὕδατι βρεχθεισα σελάσσεται, ἔσβεσε δ' αὐτὴν
> Τυτθὸν ὅτ' ὀδμήσηται ἐπιῤῥανθέντος ἐλαίου."

[4] See Hill's Theophrastus, Notes, p. 55.
[5] De Mirab. Auscult., vol. i., p. 1162.

broken into coals,[1] and are more durable; as those in a mine near Bina, and those which the river brings down; for they take fire when coals are placed upon them, and burn as long as one continues to blow them, and after they expire may again be kindled, so that they can be used for a long time, but their odor is very strong and disagreeable."[2] What place and river Theophrastus here speaks of is not known. Hill, upon what authority he does not say, finds them in Thrace, and supposes the mineral to be the same with the Thracian stone just now mentioned.[3]

Another stone, which occurred in quantity at the promontory called Erineas, "like that at Bina, emitted when burned the odor of bitumen, and the result of the combustion resembled calcined earth."[4] This was perhaps bituminous limestone, like that found near the Dead Sea.

If these minerals were neither of them bitu-

---

[1] Or, according to the conjecture of Saumaise (who for *θραύσει* would read *καύσει*), "become coals in burning."

[2] Theoph., c. 23.

[3] Hill's Theoph., Notes, p. 54. Stephanus, De Urbibus, speaks of Benna, a city of Thrace, and of Byna, a city of Crete. One or the other of these may have been the Bina of Theophrastus.

[4] Theoph., c. 27.

minous coal, there can be little doubt, however, entertained respecting that which follows; as to which Theophrastus says that "of frangible minerals, those called simply coals, because of the use they serve, are earthy, but are kindled and burned like coals. They occur in Liguria, as does amber also, and in Elis, as you go to Olympia by the mountain road. Of these, smiths make use."[1]

That mineral coal was known and even used in the time of Theophrastus these passages seem to afford sufficient proof. That it was so little used and spoken of among the ancients may be accounted for in various ways. The want of so simple a contrivance as a grate may have prevented its ordinary use as fuel, or it may have been neglected from the greater abundance of other fuel at a period of the world when so much larger a portion of Europe was still clothed in forest; besides that, in the then state of society, and of the industrial arts, and under the benignant climate of a great part of Greece, there was less occasion for fuel of any kind.

Spinus, which Theophrastus classes with com-

---

[1] Theoph., c. 28.

bustible minerals, and of which he says that, "being taken from the mine, it was cut up, and put together in the sun, when it took fire spontaneously, and especially if moistened previously,"[1] was either pyrites, or (which is more probable) an aluminous shale like that near Weymouth, or at Lyme in Dorsetshire.[2] Aristotle speaks of this stone called spinus (σπῖνος) in like manner, but somewhat less clearly than Theophrastus does.[3] We learn from him, however, that it was a Thracian mineral; and that alone, perhaps, has caused it to be by some confounded with the Thracian stone. Saumaise supposes this spinus to be mineral coal.[4]

In the case of some substances, a difference of manners and customs between ancient and modern times, or the discovery of new properties in the minerals themselves, has caused them to be used and esteemed very differently by the ancients and by us. As examples may be named asbestus and the magnet. The former is a substance full as well known to us as to the

---

[1] Theoph., c. 24.
[2] See Bakewell's Geology, c. xii., p. 195.
[3] Arist., De Mirab. Auscult., vol. i., p. 1153.
[4] Exercit. Plin., p. 179, b. D.

ancients; but by them, in consequence of their often burning the bodies of their dead, and of the religious care with which they guarded against the extinction of their sacred fire, it was applied to peculiar uses.

It probably derived its name, ἀσβέστος, that is, *unextinguished*, from its being used to form wicks for the lamps which maintained perpetual fire in many ancient temples. Such lamps Strabo and Plutarch style ἀσβέστα, unextinguished, or perpetual;[1] and Pausanias speaks of a golden lamp of this kind, made by Callimachus, an Athenian artist, for Minerva, which, "though it was kept ever burning, as well by day as night, was only once a year supplied with oil, and had a wick made of Carpasian linen, the only linen which is not consumed by fire."[2] This linen, called Carpasian from Carpasia, a town of Cyprus, was of the same kind with that described by Strabo as made of the Carystian stone, an asbestus found at Carystus in Eubœa;[3] napkins of which, he says, were, when soiled, thrown into the fire and cleansed, as others are by washing. And it was in con-

[1] See Strab., p. 396; Plut., De Def. Orac., vol. ii., p. 410, B, 411, C.

[2] Paus., lib. i., c. 26.

[3] Strab., p. 446; Brongn., Tr. Elem. de Min., i., 482.

sequence of this that the variety of asbestus suited for such use was called ἀμίαντος, amianthus, *pure, undefiled;* because, being indestructible in any ordinary fire, it was restored to its original purity and whiteness simply by casting it into the flames.[1] Where amianthus occurs, as it does in many countries, with fibres sufficiently long and flexible for that purpose, it is often now, as anciently it was, spun and woven into cloth, and has in modern times been successfully manufactured into paper, gloves, purses, ribbons, girdles, and various other things. The natives of Greenland even use it for the wicks of lamps,[2] as we have just now seen that the ancients did.

One of the ancient localities of this mineral was the island Cyprus, where it is found "of a superior quality, as flexible as silk, and perfectly white; finer, and more delicately fibrous than that of Sicily, Corsica, or Norway."[3]

There is no sufficient reason for the doubts which some have entertained, that the asbestine linen, or linum vivum, spoken of by Pliny, was of this same material; table napkins of which that author says he had seen, at an entertain-

[1] Dioscor., v., 158. [2] See Jam. Min., i., 530.

[3] Clarke's Travels, vol. iv., p. 45.

ment, blazing in the fire, and, whatever soiled them being thus burned out, rendered brighter than they could have been by water.[1] The funeral dress of kings, he adds, being made of this material, preserved their remains distinct from the ashes of the pyre. But he speaks of it as a rare and costly cloth, the material being difficult to weave by reason of the shortness of the fibre. It is true that Pliny adds some fables respecting this linen, but it is not unusual with him thus to jumble truth and falsehood.

Of amianthus he has elsewhere observed that it resembles alumen, and remains uninjured in the fire (alumini similis nihil igni deperdit);[2] copying, no doubt, Dioscorides;[3] but omitting the epithet fibrous (σχιστὴ), which his original has added to alumen, and in which only we may suppose the resemblance to have lain. Theophrastus characterizes amianthus better by comparing it to rotten wood.[4]

With respect to the magnet, its peculiar property was well known to the ancients many centuries before the Christian era, and is alluded to by Hippocrates, but merely to designate the stone, which he prescribes as remediate in a

[1] Plin., H. N., xix., 1.
[2] Ibid., xxxvi., 31.
[3] Dioscor., v., 156.
[4] De Lapid., c. 29.

certain case;[1] for it was as a remedy only, or as a charm, that the ancient Greeks used the magnet, unless we would call a use of it that strange attempt of Dinochares to construct for Ptolemy Philadelphus the vaulted roof of a temple beneath which the statue of his queen, Arsinoe, should be held suspended in the air.[2]

It has been asserted that the ancient Greeks, as far back as Aristotle, possessed some imperfect notion of the polarity of the magnet, and Majolus cites a lost work of that philosopher to prove him acquainted with this property.[3] If the fact were so, we might easily discern the origin of those other notions: of a stone called Theamedes,[4] which repelled instead of attracting iron; and of two mountains near the River Indus, one of which attracted and the other repelled that metal.[5] But there is little ground for ascribing this knowledge to the ancient Greeks, and still less reason to believe what

[1] Hippocr., Op., p. 686, l. 46. He describes it as the stone which forcibly draws iron (λίθον ἥτις τὸν σίδηρον ἁρπάζει).

[2] Plin., H. N., xxxiv., 42.

[3] As to this supposed work of Aristotle, De Lapidibus, see, besides Majoli Dies Canic., p. 566, Azuni, Dissert. sur l'Orig. de la Boussole, p. 34.

[4] Plin., H. N., xxxvi., 25; Beck., Hist. of Inv., i., 143.

[5] Plin., H. N., ii., 98.

some authors have ventured to assert, that the Egyptians, Phœnicians, and Carthaginians made use of the magnetic needle in their navigation. It seems, however, to be sufficiently proved that the polarity of the magnet and the use of the magnetic needle were known to the Chinese at a very early period, many centuries even before the Christian era.[1]

As one and the same mineral sometimes received among the ancients different names, according as it was procured by different methods, from different places, or from substances apparently unlike, so, on the other hand, things of dissimilar nature were called by the same name, merely because of some accidental agreement in color, place of origin, or use to which they were applied. Thus the name magnet was given not only to what we call the native magnet, magnetic oxyde of iron, but to a substance wholly different, and which appears to have been some variety of talc. It is highly probable that these two minerals, so different in character, were both denominated the mag-

[1] See this question examined in the Dissertation of Azuni just now cited. See Lettre à M. le Baron de Humboldt sur l'Invention de la Boussole, par M. I. Klaproth; or Silliman's Journal, vol. xl., p. 242; or Cosmos, vol. i., p. 180.

netic stone, from their being both found in a country called Magnesia; for of the five localities specified by Pliny, whence as many varieties of magnet were obtained, one is Magnesia in Thessaly, and another a city of Asia bearing the same name. And it was here, he says, a magnet was found, of a whitish color, somewhat resembling pumice, and not attracting iron;[1] which, taken in connection with what Theophrastus says of the magnet, that it was suited for turning in the lathe and of a silvery appearance,[2] leads to the inference that this magnet was talc, a certain variety of which, as it well resists the fire, and is otherwise fitted for such purpose, was employed anciently for forming culinary vessels, as it still continues to be in certain countries where it occurs in sufficient quantity and of the proper kind.[3] This mineral contains in large proportion the earth called magnesia, a name of which we may thus trace the origin; though, perhaps, a much purer form than this talc affords of the earth now called magnesia may have been sometimes designated as the magnesian stone. For when Hippocrates prescribes[4] the use of it (τῆς μαγνησίης λίθου)

---

[1] H. N., xxxvi., 25.

[2] Theoph., c. 73.

[3] Cleav. Min., p. 442.

[4] Page 543, l. 27.

as a cathartic, it seems probable that he meant the native carbonate of magnesia, which is occasionally found in an earthy, as well as in every intermediate state between that and the most compact. He certainly does not intend the magnet; as well because it is not purgative, as because he elsewhere[1] describes that differently, as the stone which draws iron, and, moreover, would have named it, not the Magnesian, but the Heraclean stone.

The story of the discovery of the magnet by one Magnes, a shepherd, on Mount Ida, who found his hob-nailed shoes and iron-pointed staff cling to the rock upon which he trod, seems to be a poetical fiction, derived by Pliny from Nicander.

The better derivation, just before suggested, is the one which Lucretius has adopted for the mineral:

> Quem magneta vocant, patrio de nomine Graii,
> Magnetum quia sit patriis in montibus ortus.[2]

Indeed, it appears that in the case of this mineral, as in that of nitre already mentioned, the ancient name, formerly applied to distinct substances, has been retained by us in connec-

---

[1] See before, p. 154.

[2] Lucr., De Nat. Rerum, vi., 808.

tion with that one to which it was least appropriate; for the name magnet or magnetic stone belonged properly to the *Magnesian* mineral just now spoken of; while the mineral by *us* called magnet was, by the ancients, more appropriately called the Heraclean stone, from Heraclea, a city of Lydia.[1] The name magnetic stone (μαγνῆτις λίθος) was given to it, as to some other minerals, simply because they occurred in Magnesia, and was extended afterward to certain substances that bore resemblance to these Magnesian minerals, without regard to their localities. Thus the magnes lapis spoken of by Pliny[2] as used in making glass was probably manganese, which had received that name from its resemblance to magnetic iron ore; for which it was mistaken, not only by Agricola and Kircher, but even at a later period.[3]

As the loadstone was sometimes confounded, in name at least, with steatite, because like that brought from Magnesia, so was it, from its be-

---

[1] See Hesych., ἡρακλεία λίθος. Plato speaks of the wondrous phenomena exhibited in the attraction of amber and the Heraclean stones (τὰ θαυμαζόμενα ἠλέκτρων περὶ τῆς ἕλξεως καὶ τῶν ἡρακλείων λίθων) in Timœo, v., 7, p. 88, edit. Tauch. See, also, Io, c. 5, a very remarkable passage.

[2] H. N., xxxvi., 66.

[3] Beck., Hist. of Inv., iv., 59.

ing found in Lydia, confounded[1] also with another wholly different Lydian mineral, the basanite, or touchstone, designated as the Lydian stone (lapis Lydius) because the best were obtained from the channel of the Tmolus, a Lydian stream. These Lydian stones were rolled pieces of silicious slate, which Theophrastus describes as resembling a smooth flat pebble, about twice as big as the largest counter (ψή-φος).[2] According to Pliny they were of moderate size, not exceeding four inches in length by two in breadth. The same substance, however, was found in large masses, and there remain some remarkable ancient statues formed of it, as those two of captive kings in the Capitol at Rome. Pliny calls this stone coticula, but observes that by some it is called the Heraclean and by others the Lydian stone.[3] In his time it was found in various other places besides the bed of the Tmolus. From what he and Theophrastus say about the mode of using it, we may infer that the ancients were capable of detecting, not only a slight degree of base

[1] Plin., H. N., xxxiii., 43. [2] De Lapid., c. 80.

[3] H. N., xxxiii., 43. Ovid, who fables the metamorphosis of Battus into this stone, calls it index. See Metam., lib. ii., v. 706.

alloy, but even when copper and silver contained a small portion of gold (χαλκὸν κατάχρυσον καὶ ἄργυρον). Pliny says they could determine how much of gold, how much of silver or of copper might be in the mass, to the smallest difference, with surprising and unfailing art. The Greeks, especially, appear, not only from what Theophrastus says, but from their frequent employment of metaphorical language derived from its use, and the many allusions to it by their writers, to have been familiarly acquainted with this mineral.[1]

But distinct things were comprehended under one name, not only because they agreed in some striking property, or as to their place of origin, but in some cases because they were applied to like uses. Thus we find that white clay or chalk, quick-lime, and the sulphate of lime were all of them occasionally called gypsum, a name which is now appropriated to the last. The term signifying an *earth that has been subjected to the action of fire*,[2] was very

[1] See Theog., v. 449; Pind., Pyth., x., 105; Bacchyl., Anal. Brunck., i., 149; Soph., Œd. Tyr., v. 492; Eurip., Med., v. 516; Theocr., Id., ii., v. 36; Plat., Gorg., c. 33; Clem. Alex., Strom., lib. i., p. 291, D.

[2] Γύψος, οἱονεὶ γηέψος τις οὖσα, ἡ ἑψηθεῖσα γῆ.—*Etym. Mag.*, p. 222.

properly applied to the last two substances, and it may have been extended to clay or chalk likewise, because that also might be used, as the others sometimes were, to whiten surfaces.

That such want of discrimination may not be imputed to gross ignorance, let us bear in mind that we are ourselves often chargeable with like inaccuracy; when, for example, we speak of crayons of black chalk, of red-lead and black-lead pencils; calling thus certain substances by the name of others to which they bear no relation, except in so far as they are applied to like use; and again, when we name differently chalk, whiting, Paris white, and Vienna white; all one and the same substance, though prepared for use in different ways, or procured from different places.

Though the name gypsum was applied, as has been said, to different things, yet it more commonly signified the sulphate of lime, which is still so called. Theophrastus and Pliny seem to have been well acquainted with the mode of preparing it, and with certain of its properties and uses; but as the latter is aware of some connection between lime and gypsum,[1]

[1] He observes (xxxvi., 59), "cognata calci res gypsum est."

so the former sometimes uses the term γύψος in the more general sense just now assigned to it of *an earth prepared by fire*, and evidently confounds it with quick-lime, when he says that its heat, upon being moistened, is surprising, and that when used in building they break it up, and, pouring water upon it, stir it about with sticks, since its heat is such that they can not with the hand.[1] The circumstance, too, he mentions of a ship laden with cloth and gypsum, which, in consequence of the water having reached the cargo, took fire and was burned,[2] is to be explained by supposing what is called gypsum on this occasion to have been lime. Epictetus speaks of a house whitewashed with gypsum, meaning, probably, lime; as do, also, Herodotus and Athenæus, when speaking of gypsum employed for a similar purpose of whitening objects.

Theophrastus distinguishes between the earthy gypsum and that prepared by burning certain stones; the former being used chiefly as a substitute for Cimolian earth in cleansing woolen garments.[3] This earthy gypsum, called

---

[1] Theoph., c. 112.
[2] Id., c. 118.
[3] Id., c. 110; Pliny, H. N., xxxv., 57.

Tymphaic from Tymphaia in Ætolia,[1] was probably such as is still found in certain places forming "beds of considerable thickness resting on other varieties of gypsum."[2] Theophrastus mentions the superior fitness of gypsum for taking impressions;[3] and it was, not long before his time, first employed for taking moulds from the human face by Lysistratus, a brother of the famous statuary Lysippus, whom Alexander distinguished by his favor.[4]

The use of gypsum as stucco in cornices, approved by Pliny,[5] is expressly forbidden by higher authority in matters of this kind, that of Vitruvius, who says "hisque (coronis) minime gypsum debet admisceri."[6] Gypsum was used, moreover, to clarify wine and preserve it from acidity; to coat vessels so as to close entirely their joints, and to pack up and preserve grapes in jars.[7]

The doubts which have been entertained whether the gypsum of the ancients was the same with ours, were occasioned, probably, by the confusion above noticed; but there can be

---

[1] Plin., H. N., iv., 3.
[2] Cleav. Min., p. 208.
[3] Theoph., c. 116.
[4] Plin., H. N., xxxv., 44.
[5] H. N., xxxvi., 59.
[6] Vitruv., vii., 3, 3.
[7] Geopon., p. 462, 483, 494.

no difference of opinion as to that gypsum, which Pliny says was ascertained to be the best, made of the lapis specularis, or what had a like foliated structure (squamamve talem habente), which was evidently selenite.[1] He adds that gypsum, when moistened, must be used immediately, for that it very soon hardens and becomes dry, "celerime coit ac siccatur."

Lapis specularis, in Pliny and other ancient authors, commonly signifies mica; but as selenite was applied to similar uses, we sometimes, as in the passage just referred to, find that to be the substance meant. The mineral called lapis specularis, from its use in windows, was furnished originally by Spain only, but in Pliny's time was found in Cyprus, Sicily, and elsewhere, though none equal to that of Spain and Cappadocia. That of Cyprus was probably selenite, of which substance Dr. Clarke saw, in that island, large masses lying along the road, as if they had been dropped by caravans, "as diaphanous as the most limpid specimens from Montmartre." The Cappadocian specular stone is spoken of by Strabo as forming large masses, which were exported.[2] Pliny de-

---

[1] Exercit. Plin., p. 184, a C; p. 771, a B, C, D, E.

[2] Strabo, p. 540.

scribes it as occurring in Italy embraced in flint (complexu silicis alligati).[1] It must have been found by the ancients of an extraordinary size, if it even approached the limit assigned by Pliny, who observes that it had never yet been obtained above five feet in length.[2]

Although Seneca speaks of the use of this substance for the purpose of admitting light in terms from which we might infer that such use was not yet very ancient in his time,[3] it seems then at least to have been very commonly applied to the same purposes for which window glass is now employed. Columella speaks of guarding plants against the cold by means of it, and says that in this way the table of Tiberius was supplied with cucumbers almost throughout the year.[4] Palladius directs the window of the oil-cellar to be closed with it.[5] Pliny speaks of protecting apple-trees with it,[6] and Martial alludes to a like liberal use of it in inclosing both apple-trees and vines.[7] Banqueting-rooms,[8] baths,[9] entire porticos[10] were

---

[1] H. N., xxxvi., 45. [2] Ibid. [3] Epist. 90.
[4] Colum., xi., 3, 52. [5] Pallad., i., 20, 1.
[6] H. N., xv., 16. [7] Mart., Ep., viii., 14–68.
[8] Sen., Nat. Quæst., iv., 13; De Prov., c. 4.
[9] Sen., Ep., 86. [10] Plin., Epist., i., 17–4, 21.

secured by means of it against the inclemency of the weather, while "the pure sun and unadulterated day" could freely enter.

> Hibernis objecta Notis specularia puros
> Admittunt soles et sine fæce diem.
>
> MART., *Ep.*, viii., 14, 2.

The glass (ὕαλος), in coffins of which Herodotus (iii., 24) says the Ethiopians called μακροϐίοι inclosed their dead, which was dug among them in great abundance and easily worked, was, of course, one or other of these specular stones.

One use of mica or of selenite that especially merits notice is mentioned by Pliny, when, speaking of the material of which bee-hives should be formed, he says many had made them of the specular stone, that they might see the bees at work within.[1] The Siberian mica, or Muscovy glass, is, as the specular stone of Spain and Cappadocia used to be, an article of commerce, and is sometimes seen in plates between three and four feet square. In Siberia, in Peru, and elsewhere, it is used in windows as a substitute for glass; to which it has even

---

[1] H. N., xxi., 47. He elsewhere speaks of Aristomachus, who spent fifty-eight years solely in studying bees; and Philiscus, who, from having passed his life in solitude occupied in the same study, was surnamed Agrius. (H. N., xi., 9.)

been preferred in certain cases, as on board ships of war, where it is not liable to be broken by the concussion which follows the discharge of cannon.

The alabaster of the ancients, and one principal use to which it was applied, have been already spoken of; but it is a substance which deserves some farther notice. That it was also denominated onyx[1] will appear less strange if we bear in mind that it was, commonly, stalagmitic carbonate of lime,[2] and composed of parallel undulating layers, often differing in color, so that both it and the silicious gem called onyx derived that name from a supposed resemblance to the human nail.[3] Pliny observes that the stone onyx, which some called alabastrites, had been thought by their ancestors to occur in Arabia only, but in his time was obtained from

---

1 It is spoken of by Dioscorides (v., 153) as λίθος ἀλαβαστρίτης ὁ καλουμένος ὄνυξ.

2 The name, however, was extended to stalactitic carbonate of lime, arragonite, and other translucent substances, of about the same degree of hardness.

3 Plin., H. N., xxxvii., 24. The colors of these parallel layers are often strongly contrasted. In Sicily is found an alabaster varied with bright red and yellow stripes; another with yellow and white stripes. At Mount Pellegrino, one of yellow and deep black. (Jam. Min., ii., 173.)

the neighborhood of Thebes in Egypt, from Damascus in Syria, and from many other places. He elsewhere[1] speaks of a city, Alabastron, in the Thebais, which probably received its name from the substance quarried there. This Theban alabaster is said to have been yellowish white, inclining to rose-red, like a variety now found at Alicant and Valencia in Spain, and at Trappani in Sicily.[2]

Mehemet Ali, previous to the year 1840, had reopened quarries of Oriental alabaster that appeared to have been worked at an ancient period, and had made and was then making liberal use of it in the construction of a splendid mosque and in the floors of his palace within the citadel of Cairo, for the columns and other decorations of a palace in his garden at Shoubra, and the erection of a handsome public fountain at Alexandria. These quarries lie between the Nile and the Red Sea, about seven leagues from Benisouef.

One of the earliest uses to which alabaster was applied was to form drinking-cups; after-

---

[1] H. N., v., 11.

[2] Jam. Min., ii., 173. Respecting the Egyptian alabaster, see Winck., Stor. d. A. del Dis., vol. ii., p. 11, and Brongn. Min., i. 217.

ward the feet of couches and of chairs were made of it; hollow vessels, of very large size; and columns, of which Cornelius Nepos had seen some thirty-two feet in length, and Pliny saw above thirty in the supper-room of Calistus, the freedman of Claudius.[1] Pliny notes its being hollowed to form unguent vessels, a purpose to which it was peculiarly adapted.[2] This use of it seems to have been widely diffused and ancient. Herodotus speaks of an alabaster of ointment (μύρου ἀλάβαστρον);[3] Cicero of "alabaster plenus unguenti;"[4] and in the New Testament we read of "an alabaster box of very precious ointment."[5] The name of the substance was, in fact, given to vessels made of other materials when applied to that use, so that we find in Theocritus "golden alabasters of Syrian ointment."[6] Propertius, too, by murrheus onyx[7] means an unguent vessel of porcelain; for we shall hereafter see that murrhina or murrhea were porcelain; and onyx, a synonym of alabaster, here indicates merely the purpose to which the vessel was applied. Hor-

---

[1] H. N., xxxvi., 12. [2] H. N., xiii., 2; xxxvi., 12.
[3] L. iii., c. 20. [4] 2, Acad. apud Nonium, 15, 17.
[5] Matt., xxvi., 7; Mark, xiv., 3; Luke, vii., 37.
[6] Idyl., xv., 114. [7] Prop., iii., 10, 22.

ace likewise speaks of "nardi parvus onyx;"[1] and Martial more than once uses the word onyx in the same sense.[2] Although the Greeks, therefore, who seldom looked abroad for the original of words they used, may, as modern lexicographers do, have fancied that *ἀλάβαστρον* was derived from *α* privative and *λαβὴ*, a handle, yet it is probable that they derived the name, together with the substance, from Arabia, which was for a long time the only source of it, and in whose language al batstraton signifies the whitish stone.[3] The onychytis of which Strabo speaks as occurring in Cappadocia was probably this same substance called onyx and alabastrites. The plates of crystal mentioned by him may have been calcareous spar, or, if the alabaster was gypseous, laminæ of selenite associated with it.

The *Lygdinus lapis*, which was found in Paros, and by many thought little inferior to alabaster for preserving unguents,[4] was the finest grained Parian marble. Anacreon, therefore,

[1] Od., iv., 12, 17. [2] Epigr., vi., 42, 14–7, 94, 1.

[3] Harris's Hist. of the Bible. The form *ἀλάβαστος*, which sometimes occurs, as in the verse of Crates (Athen., vi., 94), *Ἔπειτ' ἀλάβαστος εὐθέως ἧξεν μύρου*, may have been adopted from the mistaken derivation of the name.

[4] Pliny, H. N., xxxvi., 13.

by λυγδίνῳ τραχήλῳ[1] means a neck like that Parian marble to which Theocritus compares fine teeth,[2] and Horace the splendid beauty of Glycera:[3]

> "Urit me Glyceræ nitor,
> Splendentis, Pario marmore purius."

The Assian stone,[4] the Phrygian stone,[5] and that kind of schistos which Pliny says was called anthracites,[6] were probably aluminous slate containing more or less pyrites. With this supposition all that is said of these minerals will be found to agree. The Phrygian stone became red when burned, and was used in dyeing cloths. It and the Assian stone are mentioned by Dioscorides next before pyrites.[7] The principal use of it, according to him, and from which it derived its name, was by Phrygian dyers; and the Assian stone is characterized by a laminated structure, a saline efflorescence of a sharp taste, and its styptic properties.[8] The ἄνθος 'Ασίs λίθs mentioned by Lucian[9] was probably this saline efflorescence.

---

[1] Od., xviii., 27. [2] Idyl., vi., 30. [3] Od., i., 19, 5.
[4] H. N., xxxvi., 27. [5] Ibid., xxxvi., 36.
[6] Ibid., xxxvi., 36. [7] Lib. v., c. 141, 142.
[8] Respecting the Assian stone, see Mill, in Celsum, p. 191; and, respecting the Phrygian stone, p. 190.
[9] Lucian, v., 3, p. 654.

This Assian stone, as it is called by Pliny, who derives its name from Assos in the Troad, where it was obtained, is called by Dioscorides and Celsus[1] the Asian stone (λίθος 'Ασίος—lapis Asius), the last mentioned author appearing to derive its name from Asia.[2] All three agree in classing it with the stones which, from the property ascribed to them of consuming the bodies of the dead inclosed within them, were called sarcophagi.

The *chernites*, which Pliny,[3] after Theophrastus,[4] tells us was very like ivory, and in a coffin of which the body of Darius lay, was probably gypseus alabaster; and the *porus*, described by those authors as "resembling in color and hardness Parian marble, but particularly light, for which reason the Egyptians employed it for cornices in their best constructed buildings," was calcareous tufa.[5]

The stones called πυρομάχοι and μυλίαι by Aristotle[6] and Theophrastus,[7] pyrites and molares by Pliny,[8] were sometimes common com-

---

[1] Dioscor., v., 141; Cels., iv., 24.

[2] For he says "ex quo in Asia lapidi Asio gratia est" (iv., 24).

[3] H. N., xxxvi., 28.

[4] De Lapid., c. 15.

[5] See Strabo, p. 629.

[6] Vol. i., p. 590, 1153.

[7] De Lapid., c. 19.

[8] H. N., xxxvi., 30.

pact limestone.[1] The use of πυριμάχος in making iron, mentioned by Aristotle, agrees well with the idea of its being limestone; and Pliny,[2] speaking of various sorts of lime, approves of that made of mill-stone (utilior e molari quia est quædam pinguior natura ejus). It is probable, however, that the stones used in grinding differed in different countries, as they do at this day, and that they were often of the silicious kind. Strabo observes that the lava of Ætna, as it hardened upon cooling, was converted into mill-stone (λίθος μύλιας).[3] He speaks of the island Nisyros as furnishing abundance of mill-stones,[4] and of a promontory called Melæna, on the coast of Ionia, where there was a quarry of them.[5] It is not likely the geographer would have pointed out these two, and no other localities of mill-stone, if the substance they supplied had been one so far from scarce as common compact limestone. The name πυριμάχος may be understood as implying that the stone so called resisted well the

---

[1] Jam. Min., ii., 130. [2] H. N., xxxvi., 53.

[3] Page 629.

[4] These mill-stones of Nisyros seem to have been celebrated. See commentators on Strabo, p. 488, and Eustathius's Scholia on Dionysius Periegetes, v. 526. [5] Page 645.

fire; and Saumaise assigns a like meaning to the term pyrites when applied to mill-stone; it signifying in this case, he thinks, not a stone that gave fire on percussion, but which withstood fire in a remarkable degree.[1] Both Aristotle and Theophrastus observe that these stones might be melted, and would flow; but we may infer from what the latter says that this had been observed where they had been used in the construction of furnaces for smelting metals; so that the fact of their melting under such circumstances may well consist with the belief that they were limestone, and with the remark which Theophrastus goes on to make that "there are indeed some who think all stones may be melted except marble, which, being thoroughly burned, is converted into lime" (κονία). And this ability of limestone to resist the action of fire, when by itself exposed to it, having been observed, may suggest a probable origin of the name πυριμάχος, or fire-proof, by which it seems to have been sometimes designated.

Such a mill-stone as was called pyrites for the reason assigned by Pliny,[2] "because there

---

[1] Exercit. Plinianæ, p. 505. [2] H. N., xxxvi., 30.

was much of fire in it," must, one would suppose, have been of a silicious nature; and so, no doubt, they were in many instances. But the name pyrites, not confined to mill-stone, was applied to various minerals which gave sparks on percussion; as to the sulphuret of iron, now so called, with which pyritous copper was confounded,[1] and to flint; although this last was often, as at this day, called silex.[2]

The *geodes* and *œtites* of Pliny[3] were such hollow nodules of argillaceous oxyde of iron as are still called eagle-stone. The former was named from its embracing an earthy nucleus; the latter from its being found, says Pliny, in the nests of eagles, who were unable otherwise to hatch their young.[4] Pliny distinguishes four varieties of the ætites, differing in respect to the color, density, or substance of the kernel or nucleus contained within them.

The *hœmatites* of the ancients comprehended, besides our red hematite, several other oxydes of iron, as may be seen from Pliny's description of five varieties of it besides the mag-

---

[1] "Pyrites," says Dioscorides, "is a species of stone from which copper is melted" (v., 143).

[2] Virg., Georg., i., 135. [3] H. N., xxxvi., 32, 39.

[4] Hence its name ætites, from ἀετὸς, an eagle.

net;[1] for magnetic oxyde of iron also was classed with hæmatite;[2] but that, no doubt, because of the appearance it exhibited after having been exposed to a strong heat; for Dioscorides observes that hæmatite was likewise artificially prepared from the magnetic stone, which had been burned for a sufficient time.[3] Theophrastus describes the hæmatite as occurring frequently of a dry, squalid appearance, and resembling, according to its name, coagulated blood.[4] Another kind, he observes, was called ξανθὴ, from its being of a yellowish white. Dioscorides tells us it was sometimes found in the red ochre of Sinope.[5] Pliny, too, speaks of one hæmatites of a blood color (sanguineo colore), and another, by the Greeks called Xanthus, of a whitish yellow (e fulvo candicans).[6] He has before said it was found in mines, and, when burned, resembled cinnabar in color;[7] of

---

[1] H. N., xxxvi., 38. [2] Ib., xxxvi., 25.

[3] Dioscor., v., 144.

[4] It is compared to coagulated blood (*αἵμα πεπηγὸς*) by the Orphic poem likewise, and the reason is assigned:

> Ἐν γὰρ δὴ χρὼς αὐτὸς ἐτήτυμος αἵματος ἐστιν
> Ἐν δὲ κὰι ὕδατι ῥεῖα δαμασθεὶς ἀτρεκὲς αἷμα
> Γίγνεται——(v. 654.)

[5] Lib. v., c. 144. [6] H. N., xxxvii., 60.

[7] H. N., xxxvi., 37.

one variety, that its streak was blood-red; of another, that it sometimes resembled saffron.[1] From all which it is easily inferred that compact and ochrey red and brown oxydes of iron were included under hæmatite.

The androdamas, one of Pliny's varieties of hæmatite, which was of a black color, of remarkable weight and hardness, and attracted silver, copper, and iron, appears, when divested of its fabulous properties, to have been magnetic oxyde of iron.[2]

The schistos lapis, by burning which we find, from Dioscorides[3] and Pliny,[4] that the hæmatite was sometimes counterfeited, was probably an ochrey stone of a slaty structure, whence its name. The best was of a somewhat saffron color, friable, fissile, resembling in structure and the cohesion of its layers the fossil salt called ammoniac.[5]

Pumex signified not only what we call pumice, but other eroded cellular stones used in forming artificial grotto-work;[6] and the corresponding Greek term, κίσσηρις, seems to have been equally indefinite. Theophrastus de-

[1] H. N., xxxvi., 38.
[2] See Exercit. Plin., p. 774.
[3] Lib. v., c. 145.
[4] H. N., xxxvi., 37.
[5] Dioscor., v., 145.
[6] Plin., H. N., xxxvi., 42.

scribes two varieties widely differing in character, and neither of them what is now called pumice: one, that of the island Nisyros,[1] soft and friable, easily crushed by the hand alone into a sort of sand; the other, found in the lava of Ætna, of a dark color, dense and heavy, resembling mill-stone.[2] He alludes to the opinion of those who, as he says, thought all pumice (κίσσηρις) the product of combustion, except that which was formed from the froth of the sea. This opinion was founded upon the appearance of the substance, its associations, and the places where it commonly occurs, in the neighborhood of volcanoes. For himself, he seems inclined to think that different varieties of it may differ in their origin.[3] The stone he calls Liparæan (Λιπαραῖος), from the island Lipari, was no doubt obsidian, and he agrees with those who assign to it an igneous origin. "It is rendered porous by combustion," he observes, "and becomes like pumice" (κισσηροειδής); thus undergoing change, both as to color and density; for, previous to such action of fire, it is black, smooth, and dense. It occurs here and there imbedded in the pumice, but not con-

[1] Theoph., c. 36. [2] Id., c. 40. [3] Id., c. 34.

tinuous; just as in Milo pumice is itself, they say, enveloped in another stone, which, however, is unlike that of Lipari.[1] Mineralogists are aware that pumice and obsidian occur in Lipari intimately united and passing into each other, and that in Milo, also, pumice is abundant.[2]

Pliny speaks of obsidian as having been used for mirrors placed in walls, and reflecting shadows instead of images;[3] and Beckmann remarks on the accuracy of this description of such mirrors.[4] Obsidian was also used for ring-stones, and Pliny speaks of images and statues formed of it; but these are likely to have been of that factitious kind which he describes—a glass made in imitation of obsidian. This mineral was found in various countries, as India, Italy, and Spain; but originally in Æthiopia, by one Obsidius, according to Pliny, who in this way derives its name. But as the Greeks called it ὀψιανὸς λίθος, Saumaise[5] and Hardouin[6] prefer to derive its name, ἀπὸ τῆς ὄψεως, from its translucent nature.

Pliny describes it as of a very dark color, sometimes translucent; and an inedited Greek

[1] Theoph., c. 25.
[2] Cleav. Min., p. 306.
[3] H. N., xxxvi., 67.
[4] Hist. of Inv., iii., 185.
[5] Exercit. Plin., p. 64.
[6] Notes on Plin., v. 9, p. 782.

author, cited by Saumaise,[1] as not very black, but inclining to green.

All that is said of the native obsidian of the ancients agrees so well with what we call obsidian, and so little with any kind of marble, that it is surprising any should have supposed it to be Chian marble.

Among the stones which Pliny mentions as used for physicians' mortars, were the Etesius, the Thebaicus, the basanites and chrysites, the Tænarius, Pœnicus, and Parius lapides.[2] The lapis Etesius, which he prefers, was a species of porphyry.[3] The Thebaicus, called pyropœicilus,[4] was the red granite quarried near Syene, and thence called Syenites.

The chrysites and basanites were, probably, one and the same thing; the basanite or Lydian stone being called chrysites from its use in testing gold.[5] The Tænarius, Pœnicus, and Parius lapides, made use of for this purpose, were no doubt marbles.

---

[1] Exercit. Plin., p. 64.

[2] H. N., xxxvi., 43.

[3] Exercit. Plin., p. 776, b C.

[4] Plin., H. N., xxxvi., 13, 43.

[5] Hesych., word χρυσίτις — Exercit. Plin., p. 776. We find Pliny, too, using the term coticula in one place to denote a little mortar (xxxvi., 13), and elsewhere for the basanite or touchstone (xxxiii., 43).

The stone of Siphnus (the present Siphanto), and that found at Comum (now Como), which, being hollowed in the turning-lathe, were formed into culinary vessels, was that variety of talc called pot-stone (lapis ollaris), from the uses to which it was applied. This Siphnian stone is very particularly mentioned by Theophrastus as capable of being turned and carved because of its softness, and as used to make vessels for the table.[1] Such vessels are still in very general use in certain countries; as among the Grisons,[2] in Upper Egypt, in Greenland, and on Hudson's Bay;[3] and at Zoblitz, in Hungary, there occurs a serpentine, which is cut, turned, and polished into vessels that are distributed all over Germany.[4] To this Siphnian stone of Theophrastus and Pliny is closely related, no doubt, the Magnesian stone (μαγνῆτις λίθος) of the former author, which has been already mentioned.[5] Of that he says it might be turned in the lathe, and resembled silver, though a wholly different substance. It was, probably, a variety of talc, exhibiting a pearly, or what he calls a silvery lustre. And the

---

[1] De Lapid., c. 74. [2] Cleav. Min., p. 442.
[3] Jam. Min., i., 516; Brongn. Min., i., 487.
[4] Jam. Min., i., 510. [5] See before, page 156.

white ophites, before mentioned as a material of which hollow vessels were formed,[1] we may conjecture to have been of a like kind; especially since Pliny calls it soft, and mentions in the very next line the Siphnian stone.

Among the many varieties of whetstone which Pliny says there were, he specifies the Cretan and Lacedæmonian as requiring oil; the Naxian and Armenian, used with water; and the Cilician, with which either oil or water might be used.[2] The first mentioned were probably the novaculite, a mineral first brought into Western Europe from the Levant, and still sometimes called Turkey-stone. The others are to be referred to the various substances still applied to the same use; such as argillite, graywacke slate, mica slate, and sandstone.

The smyris (σμύρις λίθος) of Dioscorides, "a stone with which the engravers of gems polish stones,"[3] and which Hesychius describes as "a sort of sand with which the harder stones are polished," is thought to have been emery.[4]

All that Theophrastus says upon this subject is that the stone with which seals are engraved

[1] See before, page 108. [2] H. N., xxxvi., 47.

[3] Dioscorides, v., 166.

[4] See Hardouin Notes on Pliny, vol. x., p. 173.

is of the same substance with, or resembles that of whetstones (ἀκόναι), and comes from Armenia.[1]

OF MINERAL SUBSTANCES THAT WERE CLASSED WITH GEMS.

The branch of our subject in which the ancients were, perhaps, least deficient, is that relating to the stones commonly called precious. These do not in our mineralogy constitute a separate class, there being great dissimilarity among them in some important characters; but their agreement as to others; their superior weight, lustre, and hardness, together with their comparative scarceness, have, in common discourse, assigned to them a place apart from more vulgar minerals.

This consideration, it appears, led Pliny to treat of them in a separate book—the 37th and last of his great work—and though he does not even here confine himself very closely to his subject, yet has he here made several attempts at classification. In the first place, he arranges according to their several colors the more valuable gems; comprehending in this list almost

---

[1] De Lapid., c. 77.

all the stones that are recognized by us as precious, together with many that are not. We are then presented with four other lists. The first, a numerous one, contains, exclusive of varieties specified under several of the heads, no less than one hundred and thirty-five names alphabetically arranged. The next is of a few minerals which derive from parts of the body their distinctive appellations. The third is of those named after certain animals, and the last of such as are denominated from other natural objects.

In the alphabetical list, we find here and there a mineral sufficiently characterized to determine what it was; but, as to far the greater part of them, it remains wholly uncertain to what species in the mineralogical systems of the present day they would have been referred. There is, however, reason to believe that, in these lists of gems, a large proportion of the crystallized specimens in modern cabinets would have found their appropriate place and name. Many such crystallized minerals, which by their color, form, and lustre could not fail to attract attention, must undoubtedly have been known to ancient naturalists; but have passed without notice, unless they are, as is here conjectured, included in these lists. New species and new varieties must have occurred anciently, from time to time, as now; and it is of such, no doubt, our author speaks, when he says that new and nameless gems were sometimes unexpectedly discovered.

Pliny dates from Pompey's victory over Mithradates the introduction into the Roman state of a taste for pearls and Eastern gems, as also of the vessels called murrhina.[1] What these murrhina were has been a question much debated, and the controversy does not appear to be as yet decided; though Sir Wm. Gell says that "they seem at last to have been successfully traced to China. Propertius calls them Parthian;[2] and it seems certain that the porcelain of the East was called mirrha di Smyrna to as late a date as 1555." This opinion of Sir Wm. Gell is that of Dr. Clarke also,[3] was long ago maintained by Saumaise,[4] and seems more probable than any other. Two conjectures mentioned by Jameson:[5] that of

[1] H. N., xxxvii., 6, 7.

[2] Propert., iv., 5, 26. And what Propertius says of them agrees perfectly with the belief that they were porcelain:

"Murrheaque in Parthis pocula cocta focis."

[3] Travels, vol. viii., p. 151. [4] Exercit. Plin., p. 144.

[5] Min., vol. i., p. 207, 502.

Baron Veltheim, that they were figure-stone, and another, that they were concentrically-striped onyxes,[1] are undeserving our consideration, being wholly inconsistent with the greater part of what is said about murrhina by the ancients. All that Pliny remarks concerning them as of his own knowledge, will agree with the idea that they were formed of a whitish, opaque, opalescent glass, or of very fine china.[2] That they came from the East was certain; but all that was said farther about their origin seems to have been mere conjecture. Thus the Romans used cœruleum Indicum (indigo), and atramentum Indicum (Indian ink), without any knowledge of the nature of these substances;[3] and wore silk for centuries before they learned that it was not combed from the leaves of trees.[4] Pliny elsewhere speaks of murrhinum as one of the many varieties of glass that were manufactured; as a black and opaque glass, to resemble obsidian; an opaque red glass, called hæmatinon; and white, and mur-

[1] This is the opinion of Leblond, as stated by Millin, Etude des Pierres Gravées, p. 16. Corsi (Pietr. Ant., p. 166, *seq.*) labors through thirty pages to prove them to have been fluor spar.

[2] H. N., xxxvii., 8.

[3] See ante, p. 88, and Plin., H. N., xxxv., 25.

[4] Virg., Georg., ii., 121; Plin., H. N., vi., 20.

rhine, and like the hyacinth, and the sapphire, and of every other color; though the most esteemed, he says, was that of a pure transparency, and as much as possible resembling crystal.[1]

As for rock-crystal, it was the universal opinion of ancient naturalists, and the belief, indeed, almost to our own time, that it was water congealed to that hardness by long-continued and intense cold. "That it is ice is certain," says Pliny, "and hence the Greeks have given it its name."[2] This ancient notion will appear less ridiculous if we consider that, although water really converted into a solid crystalline mass by exposure to a very ordinary degree of cold resumes its fluid state when the heat of which it was deprived is again restored, yet the results of chemical analysis teach us that water in a permanently solid state constitutes a considerable proportion of many crystalline substances. Of the hydrate of magnesia, for example, it forms very near one third; and of the sulphate of soda considerably above one half. Rock-crystal is one among the very few minerals whose crystalline form Pliny has remarked. He observes that "it is not easy to

[1] H. N., xxxvi., 67.

[2] Ib., xxxvii., 9; Sen., Nat. Quæst., iii., 25.

ascertain the reason why crystal is produced in six-sided prisms (sex angulis lateribus), especially since the terminations are not uniform." That drinking-vessels were made of it proves in their lapidaries a high degree of skill; and the perfection to which the manufacture of glass had been brought in Pliny's time, may be inferred from his remark that glass vessels had approached to a wonderfully near resemblance of the crystalline; while crystals, notwithstanding, had increased, instead of diminishing in value.[1] He mentions one remarkable use of crystal in applying actual cautery. "I find," says he, "that physicians, if there are parts of the body to be burned, think that this can not be more advantageously done than by means of crystalline balls opposed to the solar rays."[2]

[1] H. N., xxxvii., 10.

[2] Ibid. He elsewhere (xxxvi., 67) speaks of hollow balls of glass, filled with water, setting fire to clothes; as does Lactantius also (De Ira Dei, c. 10); and Seneca remarks of such glass balls, that "letters, though small and indistinct, appear, when viewed through them, enlarged and more distinct" (Nat. Quæst., i., 6). This passage might be adduced as among those which favor Winckelmann's conjecture that the ancients made use of convex lenses in engraving gems. (See Stor. delle A. del D., vol. ii., p. 20, and note thereon.) This is the opinion of Natter, also, who remarks (Anc. Method of Eng. Prec. Stones, Pref., p. viii.),

This use of a crystalline lens is very ancient, if the ὕαλος of Aristophanes was, as the best commentators take it to have been, rock-crystal. Strepsiades asks Socrates if he had ever observed, among those who sell medicines, that beautiful transparent stone with which they kindle fire. "You mean crystal" (ὕαλον), says Socrates. "I do," says he. And, upon Socrates asking how he intends, by means of that, to get rid of the suit for five talents brought against him, he says that, taking this stone, he will, when the clerk is writing down the charge, stand off toward the sun and so melt out the letters.[1] The use of rock-crystal as a burning lens is also very particularly mentioned in the Lithica of Orpheus, a poem of uncertain date, though Ruhnken and Hermann would assign it to Domitian's age.[2]

The poet, having described "the bright limpid crystal, an emanation of divine fiery splendor," bids you, "if without fire you would ex-

---

"The art of engraving in gems is too difficult for a young man to produce a perfect piece, and when he arrives at a proper age to excel in it his sight begins to fail. It is therefore highly probable that the ancients made use of glasses or microscopes to supply this defect."

[1] Aristoph., Nubes, v., 768, *seqq.*

[2] Orphica, ex edit. Hermanni, p. 676.

cite a flame," to hold it above a dry torch, opposite to the beaming sun. "Immediately it will direct upon the torch a slender ray, which, touching the dry rich fuel, will produce smoke, then a little fire, and finally a bright flame; while the crystal itself, though the cause of fire, remains cold to the touch."[1] The poet alludes to the same property afterward,[2] saying of the lychnis that it could, like the crystal, without fire kindle flame.

Under the head of crystal may be introduced the mention of a stone called in the Orphic poem chrysothrix, or golden hair, of which there were two varieties: the one like crystal, the other resembling chrysolite; that is, our topaz; and both containing bright rays resembling hairs. These stones were probably quartz containing, as it sometimes does, capillary filaments of native gold,[3] or acicular crystals of some other mineral. This golden hair might be compared, or is perhaps the same, with the Venus hair-stone, which is quartz traversed by such acicular crystals of the red oxyde of titanium.[4] The enhydros of Pliny[5] was a crystal

[1] Lith., v., 170, *seqq.*
[2] Ib., v., 271.
[3] Cleav. Min., p. 241.
[4] Cleav. Min., p. 236.
[5] H. N., xxxvii., 73.

of quartz inclosing a drop of water. Claudian has made such a crystal the subject of no less than seven Latin and two Greek epigrams.

It is probable that Pliny, when speaking of the gem called adamas,[1] had in view, among other things, the diamond; but it is plain from the fables he relates of it, that this substance, "of highest value, not only among gems, but all human things, and for a long time known to kings only, and to very few of them," was unknown to him.

He has evidently confounded in his description several widely different minerals, to which, from their hardness, or their, in some respect or other, indomitable nature, the Greeks gave the name ἀδάμας, adamant. Thus steel was very frequently so called;[2] and those grains of native gold, which, when the gangue containing them was reduced to powder in a mortar, resisted the pestle and could not be comminuted by it, were called adamas.[3] Something of this sort Pollux meant by that flower of gold, or choicest gold (χρυσοῦ ἄνθος), which he calls

[1] H. N., xxxvii., 15.

[2] Αδάμας. γένος σιδήρου, Hesych., and see Stanley's Comment. on Æsch., Prom. Vinct., v. 6.

[3] Plin., Exercit., p. 757.

adamas;[1] and Plato, too, by the "branch or knot of gold (χρυσοῦ ὄζος), which, from its density, very hard, and deep-colored, was called adamas."[2] It was, no doubt, this native gold that was spoken of in the authors from whom Pliny drew when he wrote that adamas is found in gold mines; that it accompanies gold; that it seems to occur nowhere but in gold; that it is not larger than a cucumber-seed, nor unlike to it in color.

Of the six kinds he mentions, that described as occurring in India, not in gold, but bearing some resemblance to crystal, may have been the diamond; though even here it is probable that he, and those from whom he copies, mistook fine crystals of quartz for diamond, or, rather, call such crystals adamas. The description given is precisely that of a crystal of quartz in which the prism has entirely disappeared, leaving a double six-sided pyramid upon a common base.[3]

The manner in which Dionysius Periegetes characterizes adamas may lead us to suspect that he also spoke of crystals of quartz; for the diamond in its unpolished state, as known

[1] Onom., l. 7, § 99. [2] Tim., v. 7, p. 57, edit. Tauch.
[3] Plin., xxxvii., 15.

to the ancients, would hardly have been styled all-resplendent (παμφανόωντα),[1] and afterward brilliant (μαρμαίροντα).[2] The locality in the former case, too, being Scythia.

The variety of adamas which Pliny calls siderites was magnetic iron ore;[3] and the Cyprian was probably emery, or some similar substance used in engraving gems.[4]

There is a strange jumble of truth and fable in what Pliny says of the hardness of adamas, which was so unspeakable that it could not be crushed, but would split hammers and anvils used in the attempt. This invincible substance, however, which resisted the violence of fire, and the force of iron hammers, might be so far softened by steeping it in fresh warm goats' blood, as to render it possible with repeated blows to break it; though to accomplish even this the best of hammers and anvils were required. And being thus broken, it was reduced to the smallest and scarce visible particles, which, sought after by engravers, and in-

[1] Dion. Perieg., v. 318. [2] Id., v. 1119.

[3] Exercit. Plin., p. 773, 774; Jam. Min., i., 41.

[4] According to Saumaise, who would read "qui e Cypro venit" for "quod et Cyprio evenit," in Plin., xxxvii., 15.—See Exercit. Plin., p. 774, a F. Corsi (p. 271) supposes this Cyprian adamas to have been the Oriental sapphire.

closed in iron, would readily cut into the hardest substances. The localities of the several varieties of adamas were Æthiopia, India, Arabia, Macedonia, Cyprus, and Germania.

Beckmann thinks the iaspis of Orpheus[1] may have been diamond; for the poet, he remarks, "compares his iaspis to rock-crystal, and says that it kindles fire in the same manner."[2] But the learned antiquary was led astray by a false reading; for the poet compares not the iaspis, but the lychnis, to rock-crystal, and by a poetical exaggeration ascribes to it the same power of kindling fire.[3] Beckmann farther conjectures that "the iapsis in the Revelation of St. John, described as a costly, transparent, crystalline stone, was perhaps our diamond, which was afterward every where distinguished by that name."[4]

No mineralogist, who reads what Pliny says

---

[1] Lith., v., 264. [2] Beck., Hist. of Inv., iv., 238.

[3] The true reading unquestionably is *Λύχνι, σὺ δ'* instead of *Αὔχμης δ'*. (See Hermann's edition of the Orphica.) This stone (λυχνὶς) Dionysius Periegetes (v., 329) describes as altogether resembling flame (*πυρὸς φλογὶ πάμπαν ὁμοίῃ*), and it was from this fiery splendor of the stone itself, perhaps, that the Orphic poet was led to ascribe to it such virtue. See, also, Lucian, iii., 478.

[4] Beck., Hist. of Inv., iv., 239.

of the smaragdus, can fail to perceive that he classes together under that name several wholly different minerals, but what they were it is not easy to determine. Having assigned the superiority among gems to the adamas, he places next in value pearls, which also he regards as gems, and the third rank he assigns to the smaragdus. Of this he enumerates twelve kinds: first the Scythian, second the Bactrian, and third the Egyptian; and with one or other of these three our emerald, if found among the smaragdi, would probably be classed. The other kinds, he says, were found in copper mines; the best in those of Cyprus; and the characters which he ascribes to certain of them agree so well with those of malachite, that we can hardly doubt that this mineral was classed with his smaragdus. For of those smaragdi which were obtained in copper mines, some were not translucent; were of various shades of green; resembled the eyes of cats or panthers (that is, were, as modern mineralogists express themselves, chatoyant).[1] Though Pliny does not here contradict Theophrastus in any point, yet it is evident that he is compiling from some

[1] Plin., H. N., xxxvii., 18.

other source. The latter author states[1] that "of known and accessible places there are two where chiefly the smaragdus occurs: the copper mines of Cyprus, and the island over against Carthage,[2] where they are found more separate and distinct; since in Cyprus it is mined for as other mineral substances are, and in many veins where they seek for it alone. But it is rarely found of sufficient size for a seal, or ring-stone; they use it, therefore, to cement on gold; for it unites with the metal as chrysocolla does;[3] and some, indeed, think them of the same nature, since they are much alike in color; but chrysocolla is abundant in

---

[1] De Lapid., c. 49.

[2] Saumaise would read Chalcedon here instead of Carchedon (Carthage).

[3] This, though not the usual translation, is, perhaps, the best sense we can put upon a passage far from being clear. It is so understood by Kidd (Min., i., 20); and Saumaise (Exercit. Plin., p. 128, b F) seems to interpret it in the same sense with a passage which afterward occurs (c. 63), where Theophrastus speaks of the Bactrian smaragdi as being small, and used *εἰς τὰ λιθοκόλλητα*, that is, vessels of gold or silver, to the plain surface of which precious stones were firmly attached, by a plain surface, with cement, as is still practiced in Armenia, and perhaps elsewhere. This Carchedonian, or Chalcedonian emerald, Corsi (p. 257) takes to have been green feldspar, the Amazon-stone.

gold, and yet more in copper mines, while smaragdus, as has already been observed, is rare." He adds that "it appears to be formed from jasper (iaspis); for that it was said there had been found in Cyprus a stone, of which the one half was smaragdus, and the other half iaspis, which the water had not changed." He concludes by remarking that "some labor must be bestowed upon the smaragdus to give it polish, for in its natural state it has no lustre."

This mineral found in the mines of Cyprus, half emerald and half jasper, which is mentioned by Pliny also,[1] may have been arseniate of copper, accompanied, as it sometimes is, by red oxyde of the same metal.[2] The Median emeralds "were very green, and partook sometimes of the sapphire;" that is, perhaps, had the blue carbonate mingled with the green. Some were fragile; of a changeable color, resembling the green feathers of the peacock's tail, or the pigeon's neck;[3] or, in modern terms, exhibited a pavonine or columbine tarnish, such as pyritous copper often does. These same emeralds were scaly, and contained veins. The chalco-

---

[1] H. N., xxxvii., 19. [2] Cleav. Min., p. 565.

[3] "In caudis pavonum columbarumque collo plumis similes."—Plin., *H. N.*, xxxvii., 18.

smaragdus, classed with the smaragdus,[1] and described as confusedly veined with brass, was, no doubt, malachite, with pyritous copper. There are several other minerals which one familiar with their characters may see indicated by some part or other in Pliny's description of smaragdus: as chrysoberyl,[2] chrysoprase, prase,[3] plasma,[4] diallage,[5] fluor-spar,[6] green jasper, green obsidian,[7] dioptase,[8] euchroite, and euclase. The three last-named minerals have

---

[1] Plin., H. N., xxxvii., 19.

[2] Our chrysoberyl is not the mineral designated in Pliny by that name, as will appear hereafter under the head of beryl.

[3] A well-chosen specimen of prase, when cut, affords a tolerable imitation of the emerald. (Min. des Gens du Monde, p. 151.)

[4] Plasma is the prime d'émeraude of some authors. It was considered by the Romans as a gem, was cut into ornaments, and frequently engraved. (Jam. Min., i., 214.)

[5] Diallage is still called by many mineralogists smaragdite, and sometimes emeraudite.

[6] Beckmann thinks the mirror of smaragdus in which Nero viewed the gladiatorial combats (Plin., xxxvii., 16) may have been green obsidian, green jasper, or even green glass. (Hist. of Inv., iii., 177.)

[7] Green fluor-spar is called by Haüy émeraude de Carthagéne. (Tr. de Min., ii., 185.)

[8] Dioptase Haüy calls émeraude, and émeraudine. (Tr. de Min., iii., 96.)

a fine emerald color, are more or less transparent, and, in other respects, bear sufficient resemblance to the emerald to justify us in supposing that, if known to the ancients, they were comprehended under the same name. Dioptase is now found chiefly in Siberia, and, if brought anciently from the same quarter, will have been classed with the Scythian emerald, which ranked first in value. But this and other minerals occurred anciently, no doubt, in other localities than those in which they are at present found. We every now and then hear of localities that have become exhausted, and sometimes within a few years, perhaps, from the time when they were first discovered and explored. The same thing must have happened occasionally in former times; so that in some cases we may be taking fruitless pains to determine minerals that never have been seen by moderns, the localities whence only they were obtained having many ages since become exhausted or unknown.

As for the statues, obelisks, and pillars formed of emeralds of prodigious size, mentioned by Theophrastus, Pliny, and others, they were of some one or other of the several more abundant minerals that have been above suggested, or else of colored glass. Larcher thinks the pillar of emerald which Herodotus saw in the Temple of Hercules at Tyre, and which shone at night,[1] was a hollow cylinder of glass within which lamps were placed.

Theophrastus himself, speaking of this column, suggests that it may be a false emerald; "for such," says he, "there are."[2] And such are there, even at the present day, which pass for native stones. Beckmann says that a piece of glass in the monastery of Reichenau, seven inches long, and weighing twenty-eight pounds, and a large cup at Genoa, which is, however, full of flaws, are given out to be emeralds, even to the present time.[3]

Some have supposed that our emerald ought not to be reckoned among the many varieties of smaragdi mentioned by the ancients. Dutens doubts if it was known to them; and from the researches, and the positive assertion of Tavernier, it appears, at least, that no locality of emerald is known in Asia or its islands.[4] But "that emeralds were known in Europe before the discovery of America is proved by the em-

---

[1] Herod., ii., 44.

[2] De Lapid., c. 45.

[3] Hist. of Inv., iii., 189.

[4] Brongn., Tr. Elem. de Min., p. 418.

erald that was in the mitre of Pope Julius the Second, and by the necklace of antique emeralds found in Pompeii, and seen by Mr. Hawkins."[1] The Egyptian emeralds, mentioned by Pliny were, in all probability, derived from the emerald mines of Gebel Zabára, in Upper Egypt, which Bruce speaks of, and which were reopened in 1818 by Mehemet Ali, but without success.

With respect to the beryl (beryllus), Pliny observes that "many regard it as of the same, or certainly of a like nature with the emerald."[2] He seems to think its crystalline form due to the lapidary's art; but adds, that some suppose them to be naturally of that shape.[3]

The best were those of a pure sea-green (qui viriditatem puri maris imitantur), our aquamarina, beril aigue-marine.[4] The next in esteem were called chrysoberyl, and are somewhat vaguely described as "paulo pallidiores, sed in aureum colorem exeunte fulgore." This

[1] Clarke's Travels, vol. viii., p. 150, note.

[2] H. N. xxxvii., 20.

[3] "Poliuntur omnes sexangula figura artificum ingeniis —quidam et angulosos putant statim nasci."

[4] Dionysius Periegetes calls it *ὑγρῆς βηρύλλου γλαυκὴν λίθον*, that is, according to Eustathius, sea-green; and afterward merely *βηρύλλου γλαυκὴν λίθον*.—v. 1012, 1119.

was probably the yellow emerald, such as occurs in Auvergne, or at Ackworth in New Hampshire. The third was called chrysoprase; and would seem to have been, in fact, as Pliny says some considered it, a mineral *proprii generis*, different from the beryl. It resembled in color the juice of the leek, but with somewhat of a golden tinge, and hence its name.[1] Although we are uncertain as to the mineral here described, yet it is not improbable that it was the same now called chrysoprase, and to which Lehman was the first in modern times who gave the ancient name.[2] The fourth variety of beryl was of a color approaching the hyacinth; the fifth of a wax, and the sixth of an olive color. The last variety spoken of by Pliny resembled crystal, but contained hairy threads and impurities (crystallo fere similes—capillamenta habent sordesque). These were probably such crystals of quartz as are often found, rendered partly opaque by chlorite, or penetrated by capillary crystals of epidote, actinolite, or other minerals. Pliny observes that the Indians stained rock-crystal in such a way as to counterfeit other gems, and especially the beryl.[3]

---

[1] Plin., H. N., xxxvii., 33, 34. [2] Jam. Min., i., 202.
[3] H. N., xxxvii., 20.

The opal (opalus) of Pliny is too well characterized, and its peculiar lustre, or opalescence, too accurately described by him, to leave any doubt that it was what we call precious opal. He first speaks of it in the order in which the great value ascribed to it entitled it to rank: in the first class of gems.[1] Afterward, when speaking of gems according to their respective colors, he introduces it again, as taking the lead among white gems (dux candidarum), under the name of pæderos; by which name he had before said it was, because of its extraordinary beauty, more generally distinguished.

Pliny is not the only one among the ancients, as Jameson supposed, who makes mention of this gem. The Orphic poem commends the beauty of the *ὀπάλλιος*, and evidently alludes to its other name, pæderos (*παιδέρως*), in saying that it has the delicate complexion of a lovely youth (*ἱμερτοῦ τέρενα χρόα παιδός*).[2] This gem, also, Pliny says, the Indians so well imitated in glass that the counterfeit could hardly be detected. The opal was, perhaps, too highly valued to be frequently engraved. There are very few engraved specimens of this

[1] H. N., xxxvii., 21.

[2] Orph., Lith., v. 280.

mineral preserved in collections;[1] but that it sometimes was used as a ring-stone, we learn from the story Pliny tells of a senator named Nonius, who, possessing an opal valued at twenty thousand sesterces, which Antony coveted, was proscribed in consequence, and fled, saving of his whole fortune this ring alone (e fortunis suis omnibus anulum abstulit secum).

The sardonyx (sardonyches), mentioned by Pliny next after opal, as holding the next rank, was evidently the same stone with that now so called; but under the same denomination seem to have been comprehended other varieties of chalcedony, and especially the carnelian which Werner calls sardonyx, whose colors are in alternate bands of red and white, and, when the stone is cut in certain directions, resemble the flesh seen through the finger nail.[2] The first Roman who sealed with a sardonyx was the elder Scipio Africanus, from whose time this sort of gem was much used for that purpose, it being almost the only one which left a fair impression and brought away with it no portion of the wax. This gem was most approved

---

[1] Jam. Min., i., 233.

[2] "Velut carnibus ungue hominis imposito."—PLINY, xxxvii., 23.

when it exhibited distinct colors and bands well defined. The localities mentioned by Pliny are India, Arabia, and Armenia.

The gem onyx seems to have comprehended several varieties of agate; as the onyx-agate, and the eyed-agate.[1] The name, derived from a certain resemblance to the human nail, was applied originally to the calcareous alabaster before mentioned; the substance meant whenever unguent vessels are spoken of as made of onyx.[2] From that it was afterward derived to the gem, or ring-stone onyx.[3]

The sard (sarda) of Pliny is what we call carnelian. By remarking that it was first found at Sardes, he means, probably, to suggest the origin of its name; which others prefer to derive from Sardinia, where Kircher says that very good ones are obtained.[4] Epiphanius says it received its name from some resemblance which it bore to the fish called sardine (σαρδίῳ ἰχθύϊ τεταριχευμένῳ).[5]

1 "Cingentibus candidis venis oculi modo."—PLIN., *H. N.*, xxxvii., 24.

2 See before, p. 167.

3 Plin., H. N., xxxvii., 24. Used in this latter sense the word is of the feminine gender, as in the former it is masculine; the word gemma being understood in the one case, and lapis in the other. (Plin., ubi supra et Exercit. Plin., p. 393, 396.)

4 Mund. Subterr., lib. viii., p. 81.

5 Epiph., de 12 Gemmis, p. 22.

The best sards had been obtained from near Babylon, in working certain stone quarries, where it was found enveloped in the rock; but that locality, Pliny says, had failed. It was, however, a common gem, and occurred in many other places. There was no one, Pliny says, more frequently employed among the ancients; and by referring to the plays of Menander and Philemon, to confirm his assertion, he gives us to understand whom he means to designate as ancients. He speaks of one of the three kinds brought from India as being underlaid with silver foil when set; and of another kind from Egypt, under which gold foil was laid. The favor this gem enjoyed as a ring-stone was in consequence of its making, like the sardonyx, a clean impression, and bringing away no portion of the wax.

Under the head of glowing gems (ardentes gemmæ) Pliny classes a considerable number, and assigns the first rank among them to carbunculi. The carbunculus of Pliny is said by some to have been the ruby, while others regard it as the garnet; and both opinions, probably, are right. The Latin name seems, like our term ruby, to have been applied to very different minerals, and may have comprehended the red sapphire, or Oriental ruby; the spinelle ruby; the red topaz, or Brazilian ruby; the Bohemian ruby (a variety of red quartz); red fluate of lime, called the false ruby; together with several varieties of garnet.

Those carbuncles which Pliny calls Alabandic, because they were cut and polished at Alabanda,[1] were precious garnets, still called by some mineralogists Alabandines, or Alamandines. What he afterward says of Alabandic carbuncles, which were darker colored and rougher than others, may be explained by supposing that near Alabanda both precious and common garnets were obtained.

Those Indian carbuncles which he describes as "non claros, ac plerumque sordidos, ac semper fulgoris horridi," and which had been hollowed into vessels that would hold a pint (sextarii unius mensuram), were common garnets; which are sometimes found of considerable size, as large even as a child's head.[2] It is evident that there are still other minerals besides these among the carbunculi; but what those minerals may be it is not easy to determine.

---

[1] "Alabandicos in Orthosia caute nascentes, sed qui perficiantur Alabandis."—*H. N.*, xxxvii., 25.

[2] Cleav. Min., p. 367.

Pliny observes that these gems were so easily imitated in glass, something being placed underneath to improve its lustre, that to distinguish the false stones from the true was very difficult; though, like other factitious gems, they might be detected by the lapidary's wheel, their greater softness and fragility, and their inferior weight.

Theophrastus uses the name *ἄνθραξ* (carbunculus) in a somewhat more restricted sense. Having spoken of combustible minerals, he goes on to point out some that are wholly incombustible; and first the carbuncle (*ἄνθραξ*), which, he observes, is engraved for seals, and describes as of a red color, and when exposed to the rays of the sun resembling a burning coal. It may be called very precious, he remarks, since a very small one sells for forty pieces of gold. This kind was brought from Carthage and Marseilles. Another kind, alike incombustible, occurred near Miletus, angular, and sometimes hexagonal, which also was called carbuncle (*ἄνθραξ*).[1] The *ἀνθράκιον*, which he afterward describes,[2] probably comprehended some varieties of common garnet, and perhaps schorl.

---

[1] Theoph., c. 31, 32.

[2] Id., c. 61.

The anthracites, described by Pliny among these glowing gems,[1] may be micaceous oxyde, or scaly red oxyde of iron. The latter is sometimes seen in cavities of the red hæmatite,[2] between which and anthracites there seems to have existed some connection.[3] By its being "carbonibus similes," Pliny must mean that it possessed a fiery lustre, a glow resembling that of coals, and which he presently describes as "igneus color." That, when thrown into the fire, these anthracites "velut intermortuæ extinguuntur, contra, aquis perfusæ exardescunt," means, probably, that their lustre is destroyed by casting them into the fire, but improved by sprinkling them with water; which is true.

The sandaresus and sandaster[4] were perhaps varieties of aventurine quartz, which occurs of all the colors ascribed by Pliny to these stones: red, as they probably were, being classed with carbunculi; yellowish brown, or resembling smoky topaz (fumido chrysolitho similes);[5] yellow, or "mali coloris;" and greenish, or with

[1] H. N., xxxvii., 27. [2] Cleav. Min., p. 601.

[3] See Plin., xxxvi., 38, and Dalecamp's notes thereon. Agricola makes the anthracites a species of hæmatite.

[4] Plin., H. N., xxxvii., 28.

[5] The chrysolithus of Pliny was our topaz, as will appear hereafter under the proper head.

the color "olei viridis." The transfulgent drops of gold, seen never on the surface, but always in the body of the mineral, were what mineralogists now describe as "brilliant points or spangles, which shine with a silver or golden lustre, and seem to be produced by the reflection of light from numerous fissures; or from disseminated plates of mica; or, perhaps, from laminæ of quartz interspersed through the mass."

The lychnis, so called from some resemblance to the flame of a lamp,[1] or from its peculiar lustre when viewed by lamplight, was of several kinds, distinguished by their colors; as pale-red, violet-red, crimson-red. In which respect, and as to the localities whence it was obtained, it agrees well with the rubellite. But the chief point of resemblance to this mineral is its possession of electric properties, observed by Pliny, who says that, when heated in the sun or by friction, it attracts chaff and leaves of paper (chartarum folia).[2] Nor, if there should be, otherwise, sufficient ground for believing lychnis to be rubellite, need we be staggered

[1] Plin., H. N., xxxvii., 29.

[2] This property led Beckmann to suppose it might be tourmaline. (Hist. of Inv., i., 144.)

by what our author, in the following chapter, says—that drinking-vessels had been made of lychnis, and of the mineral he is there speaking of, carchedonius; for in the very next sentence he remarks that "all these gems obstinately resist the graver's tool;" which seems wholly inconsistent with their being hollowed into drinking-vessels. They may, perhaps, have been used, as other gems were, in ornamenting such vessels, which were thence called *διάλιθα*, or *λιθοκόλλητα*.

The carchedonius, which occurred in the mountains of the Nasamonæ, and was taken thence to Carthage, from which city it derived its name, is supposed by some commentators on Pliny[1] to have been what we call chalcedony, but the characters assigned to it by Pliny do not warrant such opinion. Like the Carthaginian carbuncles (carbunculi Carchedonii) spoken of before, it had its name from Carthage (*Καρχηδὼν*), but it seems to have been a different mineral;[2] and, from its possessing electricity, though in slighter degree than the lychnis, from its brittleness, and its resemblance to expiring coals (in blackness, probably), we might

[1] Hardouin, in Plin., vol. x., p. 88; Salm., Plin. Exercit., p. 270, a F.

[2] Plin. Exercit., p. 270, a D.

not unwarrantably infer that it was schorl. There is, indeed, one argument against this supposition, and in favor rather of its being the above-mentioned variety of carbunculus, which is, that Pliny classes it with the gems he styles ardentes;[1] and which, after having spoken of sardonyx, he considers, by way of digression, before treating of the sarda, because they all agreed in certain particulars as to which they differed from the sard. They were all of them engraved with much difficulty, and, when used as signets, retained and brought away with them a portion of the wax.

The topaz (topazius), which Pliny speaks of as a gem of a peculiar green color (suo virens genere), which even in his time retained a high value, but was upon its first discovery preferred before all others, is supposed by some to have been the same with our chrysolite;[2] as his chrysolite is considered by Werner and others to have been the stone which we call topaz.

This name is derived from that of the island Topazos, in the Red Sea, whence this mineral was originally brought. Juba[3] derives the name of the island itself from τοπάζω, *to conjec-*

[1] H. N., xxxvii., 24. [2] Jam. Min., i., 48.
[3] Cited by Plin., xxxvii., 32.

ture; it being often hid in mist, and mariners, consequently, at a loss to find it.[1] Pliny styles his topaz largest of gems (amplissima gemmarum), and speaks of a statue of Arsinoe, wife of Ptolemy Philadelphus, made of it, four cubits high; which seems wholly inconsistent with its being chrysolite; although a variety of this mineral, called olivine, has been found in masses of considerable size.

The extraordinary dimensions ascribed to Pliny's topaz have led some to think it was a variety of jasper or of agate. Pliny says that it alone, of the nobler gems, could be touched by the file; and that recent authors distinguished two kinds of it: one having the color of the leek, and therefore called prasoïdes; the other, more inclined to a golden color, styled chrysopteros, and resembling chrysoprase.

Pliny's whole description of the topaz is, perhaps, as applicable to the minerals which we call prase and chrysoprase as to any that we know.

Bruce mentions an island in the Red Sea called Jibbel Seberget, or the Mountain of Emeralds; but says the substance he there met

[1] For a strange account of this island and the topaz it produced, see Diod. Sic., lib. iii., c. 38.

with was little harder than glass; he conjectures that this green, pellucid, crystalline mineral, as he describes it, was classed with the smaragdi of the ancients.[1] Kidd asks, May not this have been chrysolite, and this island the Topaz island of Pliny?[2] But later researches have shown that the substance in question is green-colored fluor-spar.[3] Pliny, after observing that this topaz "solus nobilium limam sentit," adds that the rest are polished *naxiis cotibus;* and he elsewhere states that *naxium* has long been preferred to any other substance for polishing marble statues, for engraving and polishing gems.[4] By naxiæ cotes and naxium he, no doubt, means the emery of the island Naxos.

The color and other characters, as far as they are given, of callaïs, agree tolerably well with the belief that it was what we call turquoise, a stone of which there exist in cabinets many ancient engraved specimens.[5] Pliny says the callaïs was of a pale green (e viridi pallens); that its best color was that of emerald; that it occurred in roundish masses (oculi figura extuberans), and in regions of Asia which are the

---

[1] Bruce's Travels, vol. i., c. 9. [2] Min., i., 121.
[3] Russel's Egypt, p. 431. [4] H. N., xxxvi., 7.
[5] Millin, Étude des Pierres Gravées, p. 18.

same with those whence the turquoise is now brought.[1] The story Pliny tells of its being obtained, by means of slings, from lofty and inaccessible rocks, to which it feebly adhered, projecting above their surface, we may be permitted to regard as a fable, not to be taken into consideration.

Of this gem, also, he observes that there was no one more perfectly imitated in glass.

Prasius, a green stone, of inferior rank,[2] was probably our prase; and that second kind of it, which was horrid with spots of blood; and the third kind, distinguished by three light-colored stripes, were two varieties of heliotrope. Some, says Pliny, prefer to these the chrysoprase, which also resembles in color the juice of the leek, but from that of topaz inclines to that of gold. To this chrysoprase, which he has before spoken of as a variety of beryl, he ascribes such dimensions that cups (cymbia) were sometimes formed of it. It is probable, as has already been observed,[3] that this was the same mineral as the stone of Kosemütz, which Lehman first called chrysoprase.

The prasites of Theophrastus,[4] of which he

[1] H. N., xxxvii., 33.
[2] Plin., H. N., xxxvii., 34.
[3] Page 202.
[4] De Lapid., c. 65.

merely observes that it was somewhat like verdigris in color (ἰώδης τῇ χρόᾳ), was either the prasius or the chrysoprase of Pliny.

The Nilion, so called from the River Nile, on the banks of which it occurred in Æthiopia, without lustre, dull, of the color of smoky topaz, and sometimes of honey, was, perhaps, jasper.

The molochites, malachite, so called from μολόχη or μαλάχη, mallows, which it resembled in color, is described as opaque, of a deeper and duller green than the smaragdus. We may, nevertheless, regard it as the mineral still called malachite, and which, as we have seen, Pliny classes with smaragdi.[1] He may here again introduce it, under a different name, among the green minerals he is now describing, because he finds it mentioned in the authors now before him; for he often looks no farther than the source from which at the time he happens to be compiling.

"Iaspis," says Pliny, "is green, and often is translucent." What we call jasper is of almost every color, and is opaque. But still, the ancient iaspis may have comprehended certain

---

[1] See before, page 195.

varieties of green jasper; and since agate and jasper are closely connected, and pass into each other, it is probable there were varieties of agate, also, classed under the same head. The Orphic poem, speaking of the various colors of the agate, says, "You recognize therein the glassy jasper, the blood-red sard, and the brilliant emerald."[1] The iaspis found in Persia, and described as resembling in color an autumnal morning sky, and hence called aërizusa, is thought by some to have been our turquoise; but Beckmann is decidedly opposed to this opinion, and supposes it to have been blue jasper.[2] Jameson may say with truth, "We are ignorant of the particular stone denominated jasper by the ancients;" for certainly there is no one stone to which the description of jasper could be applied; but in this case, as in others, it is evident that several different minerals were comprehended under a single name. One variety specified by Pliny, "stellata rutilis punctis," resembled a jasper, not uncommon at this day, of a dark green color, with red spots, and like heliotrope, except that it is not translu-

[1] Lith., v. 607.

[2] See Salm., Exercit. Plin., p. 143, a C; Beck., Hist. of Inv., ii., 322.

cent. All the East, Pliny says, wore as amulets that variety of jasper which resembled emerald, and had a white stripe running through the midst of it, and was therefore styled grammatias; as that which had several such stripes was called polygrammos. Here, again, he describes what we may easily recognize as striped or ribbon jasper, a mineral which abounds in many countries.

Pliny classes with jasper certain stones which he calls signets (sphragides), because they sealed so well. The better sort, he says, were set open (à jour), in such way that the gold merely embraced the margin of the stone.[1] He mentions, as though it were an unusual magnitude for this mineral, that he had seen an image of Nero in jasper, fifteen inches high. We find that this stone, also, as many others already spoken of, was imitated in glass; but the counterfeit was easily detected from the difference in lustre.

Dioscorides briefly characterizes several varieties of jasper:[2] one resembling the emerald; one somewhat crystalline, resembling phlegm;

[1] "Funda clauduntur patentes, nec præterquam margines auro amplectente."—*H. N.*, xxxvii., 37.

[2] Lib. v., c. 160.

one that has bright veins of white running through it; one sky-blue; one smoky; one resembling turpentine, but in what respect he does not mention; and one like the callaïs, which we have seen was of a light green color.[1] All of them, he says, were worn as amulets; and its virtue as such is mentioned by Dionysius Periegetes also, who alludes to it in three different places,[2] calling it sky-blue (ἠερόεσσαν), sea-green (for so ὑδατοέσσαν is here to be interpreted), and green translucent (χλωρὰ διαυγάζουσαν). The Orphic poem also styles it green (ἐαρόχροον). The only epithets that Pliny adds to those already noticed are purple and purple passing into cœrulean. Among the localities of the mineral he mentions Chalcedon as furnishing an opaque variety.

It will be found that chalcedony and jasper, which are often associated with each other, and which are both of them, but especially the former, sometimes found united with limpid and amethystine quartz,[3] will answer tolerably well to all these descriptions of iaspis. And to this same head of iaspis it is highly probable that nephrite, and other varieties of jade, were anciently referred.

---

[1] See before, page 214. [2] Verses 724, 782, 1120.

[3] Cleav. Min , p. 248.

We find that Pliny again introduces here as a gem, in the order in which we are now following him, the cyanus, which, under the name of cœruleum, he had before treated of as a pigment. This may be accounted for from the occasional occurrence of fine crystallized specimens of the blue carbonate of copper; which would naturally, according to his mode of speaking, be reckoned among gems. But it is evident that he here applies to his gem cyanus what Theophrastus says[1] of the Egyptian factitious cyanus before mentioned.[2] The golden dust which this gem cyanus sometimes contained was probably pyritous copper.[3]

The sapphire of the ancients, described by Theophrastus as sprinkled with gold (χρυσιπάστος),[4] and in which Pliny says gold sparkles (scintillat),[5] is agreed by all to have been our lapis lazuli.[6] The name is Hebrew, and occurs

---

[1] De Lapid., c. 98. [2] See before, p. 86.

[3] The cyanus of Homer, used in Achilles' shield, is, with much reason, thought by Goguet (Orig. des Lois, vol. ii., p. 162) to have been polished steel colored blue by tempering.

[4] De Lapid., c. 43.

[5] H. N., xxxiii., 21. Dionysius Periegetes also fixes upon this same character, and speaks of rocks in India which contain a beautiful vein of golden and azure sapphire:

"Χρυσείης κυανῆς τε καλὴν πλάκα σαπφείροιο."

[6] Jam. Min., i., 341; Beck., Hist. of Inv., ii., 322.

repeatedly in the Old Testament, applied to the same substance.[1] What the ancients mistook for gold was the iron pyrites often disseminated in this mineral, and forming a feature in its external character, upon which, under their mistake, they were inclined to lay much stress. It is evident, however, that other minerals besides lapis lazuli were included under the name sapphire. Pliny speaks of purple sapphires, of which the best, he says, are the Median.[2]

In the case of this mineral, as of the last-mentioned, cyanus, the distinction of male and female was adopted, those of an azure color being considered males.

Pliny, it is true, says that sapphires nowhere occurred translucent; we might otherwise suppose that the Oriental sapphire, wherever known, was included under the name; as also blue quartz, and blue fluor-spar, which are now sometimes called by jewelers false sapphire.

---

[1] Jam. Min., ii., 324.

[2] H. N., xxxvii., 39. Whatever the purple sapphire may have been, it was at a later period more highly valued than the azure stone. Epiphanius observes that there were many kinds of sapphire, and that the one called royal (βασιλικὸς), spotted with gold (χρυσόστιγης,) was less admired than that altogether purple (διόλου πορφυρίζων), which occurred in India. (De 12 Gemmis, p. 227.)

There are in collections many engraved specimens of lapis lazuli, regarded as antiques,[1] but no engraved sapphires that are not of modern date.[2]

The first rank among purple or violet colored gems is assigned by Pliny to the Indian amethyst, which may, possibly, have been the violet-colored sapphire, or Oriental amethyst. Those which he describes as easily engraved (scalpturis faciles) may have been the violet-colored fluor-spar, now called false amethyst;[3] and the variety of quartz which is now commonly styled amethyst is well described by him as that fifth kind which approaches crystal; the purple vanishing and fading into white (ad viciniam crystalli descendit albicante purpuræ defectu). Some mineralogists think the amethyst of the ancients to have been what we call garnet, but there seems little in its description resembling the garnet, except that one kind of it approached the hyacinth in color, as Pliny and Epiphanius[4] observe; that is, had a very

---

[1] Beck., Hist. of Inv., iii., 322. [2] Jam. Min., i., 37.

[3] The ancients, however, very frequently engraved upon amethystine quartz; and there are, in the Royal Library at Paris, many fine engraved gems of this mineral. (Jam. Min., i., 148.)

[4] De 12 Gemmis, p. 229.

strong shade of red; and so sometimes has our amethyst. The way in which it is characterized by Dionysius Periegetes, as in a slight degree purple,[1] suits perfectly with our amethyst, but in no respect with garnet; and we see our amethyst plainly indicated in one of the reasons assigned by Pliny for its name[2]—that it does not reach the color of wine, but first fades into violet. He afterward suggests another, which is the more common derivation; saying that the Magi falsely assured that these gems were preservative against intoxication, whence their name.

The Indian amethysts are described as possessing "absolutum felicis purpuræ colorem;" and the artists who imitated this gem endeavored especially to produce this tint. Theophrastus twice mentions the amethyst (*αμέθυσον*),[3] but not in such a way as to determine it; classing it in one place with crystal, as diaphanous, and afterward observing that it is wine-colored (*οἰνωπὸν τῇ χρόᾳ*)

The ancient hyacinth is thought by Saumaise to have been our ruby, which the Per-

---

[1] "*Καὶ γλυκερὴν ἀμέθυστον ὑπηρέμα πορφυρέουσαν*" (v. 1122).

[2] *Αμέθυστος*, from *α* and *μέθυ*.   [3] De Lapid., c. 54, 57.

sians and Arabians still call *yacut;* a name derived from ὑάκινθος, hyacinth. This name may, however, have been used with as little discrimination as that of ruby is at present, to designate several very different minerals, and among them may be some that are still called hyacinth; as several varieties of zircon, and the hyacinth of Compostella, a red ferruginous quartz. Jameson enumerates seven different minerals besides zircon, to which the name hyacinth has been applied; and he appears to think that the ancient hyacinth was either amethyst or sapphire. All that Pliny says about this gem, from which any aid toward determining it can be derived, amounts to this, that it was related to, but differed greatly from, the amethyst, in that the lustre and brilliant violet of the amethyst were much fainter and feebler in the hyacinth;[1] which certainly does not well agree with the idea of hyacinth being our ruby, but favors, rather, the opinion before noticed, that the hyacinth of Pliny was our amethyst, and the ancient amethyst our garnet.

The chrysolithus of Pliny[2] was, as Werner thinks, the stone which we call topaz. But the

[1] "Ille emicans in amethysto fulgor violaceus dilutus est in hyacintho."

[2] Pliny, xxxvii., 42.

ancient name appears to have been applied somewhat loosely, as the modern is, to a great variety of minerals. The chrysolites obtained from Æthiopia were "aureo fulgore translucentes;" but to these were preferred the Indian, which may have been the yellow sapphire or Oriental topaz. The best were set open (à jour).[1] Underneath others a foil of brass was laid. Those were called chryselectri whose color approached to that of amber (electrum). Those of Pontus might be distinguished by their lightness. They were, perhaps, yellow quartz, the Bohemian topaz; or yellow fluor-spar, the false topaz; whose specific gravities are to that of the Oriental topaz as three and four respectively to five.

The chrysolite obtained in Spain, from the same locality with rock-crystal, we may suppose was yellow quartz. Such as had a white vein running through them, called leucochrysi, were probably agate; yellow quartz, with a vein of chalcedony; and the capniæ[2] we may translate smoke topaz. Some resembled glass of a bright saffron color; and those made of glass could not be distinguished by the sight,

[1] "Funda includuntur perspicuæ."

[2] A name derived from καπνὸς, smoke.

but might be detected by the touch (of the tongue, no doubt), as being warmer.[1] Other varieties were melichrysos and xanthus, both from India: the former a hard, brittle, translucent gem, of a bright golden or honey-yellow color;[2] the latter a very common gem in India, and afterward described as of a whitish yellow.[3]

The asteria of Pliny[4] is that variety of opal called girasole, from its reflecting a reddish light when turned toward the sun. Pliny describes with peculiar felicity of language the manner of this reflection.[5] This gem came from India and from Carmania, those of the latter country being preferred. Pliny describes it as difficult to engrave; the difficulty arising, probably, not from its hardness, but from the numerous minute fissures which traverse opal in all directions, and to which it is supposed to owe the playful variation of its colors.

---

[1] "Tactus deprehendit, tepidior in vitreis."—PLINY, *H. N.*, xxxvii., 44.

[2] "Veluti per aurum sincero melle translucens."

[3] "E fulvo candicans."

[4] H. N., xxxvii., 47.

[5] "Inclusam lucem pupillæ modo quandam continet, ac transfundit cum inclinatione, velut intus ambulantem ex alio atque alio loco reddens, eademque contraria Soli regerens candicantes radios, unde nomen invenit."

The gem called astrios by Pliny, which occurred in India and on the shores of Pallene, but best in Carmania, and which, "nearly related to crystal, shines from a point within it like a star with the brightness of the full moon,"[1] we might, but for the extreme rarity of that stone, conjecture to have been asteriated sapphire. Some crystals of quartz, which exhibit irised colors,.and reflections of light from their interior, might answer to Pliny's description tolerably well; but far the most probable conjecture, as regards this stone, at least, is that of Werner, who supposes the moon-stone of Ceylon, a variety of adularia, to be the ὑαλοειδὴς[2] of Theophrastus, the asteria, the astrios, and the androdamas of Pliny.[3] Dionysius Periegetes, when speaking of Pallene, one of Pliny's localities for this gem, says, "There is produced the beautiful stone asterius, shining like a star."[4]

The astroites[5] does not seem to be determ-

[1] "Intus a centro seu stella lucet fulgore lunæ plenæ."—*H. N.*, xxxvii., 48.

[2] Of which Theophrastus says nothing more than that it both reflects and transmits light. (De Lapid., c. 54.)

[3] Jam. Min., i., 362.

[4] Verse 327.

[5] Pliny, H. N., xxxvii., 49. Saumaise would read this name astriotes. (Exercit. Plin., p. 533, b E.)

ined, though Cardan speaks of it as of something known.[1] We might, perhaps, be justified in supposing it to have been an organic fossil, the asteria, or stella marina; or, since that is a very rare fossil,[2] perhaps a madreporite.

Of the astrobolos Pliny merely says that it resembled the eyes of a fish, and had in the sunshine a white lustre. These characters would agree well with apophyllite, called fischaugenstein by Werner, ichthyophthalme and ichthyophthalmite by other mineralogists, from the resemblance of its peculiar pearly lustre to that of the eyes of a fish. But astrobolos is more likely to have been that variety of adularia called fish's eye.

The ceraunia, which Pliny now considers separately, appears to be the same mineral he has just before mentioned as a variety of what he calls astrios. The description he gives of it would, as far as it goes, be a very good one of a transparent crystal of adularia, exhibiting, as such crystal sometimes does, a bluish-white reflection from a movable spot in its interior. This serves to confirm Werner's conjecture that

---

[1] See Dalecamp on Plin., xxxvii., 49.

[2] See Park., Org. Rem., vol. iii., p. 1.

astrios was the variety of adularia which is called moon-stone.

Those other two kinds of ceraunia, the one black, the other reddish, and in shape resembling axes, seem to have been something wholly different from the one just mentioned. They were, perhaps, varieties of jasper or of the axe-stone.[1]

That fourth kind which was found only where lightning had fallen would seem to have been an aërolite.[2]

Some varieties of the ceraunia, however, were so called from a superstitious belief that they were a protection, to those who carried them about their persons, against the effects of lightning.[3] And even to the aërolite, which was called ceraunius, as Marbodæus asserts,[4] because it was thought to be found in those places only which had been struck by the thun-

---

[1] See Cleav. Min., p. 269, 340.

[2] "Ventorum rabie cum turbidus æstuat aër,
Cum tonat horrendum, cum fulgurat igneus æther,
Nubibus elisus, cœlo cadit ille lapillus,
Cujus apud Græcos extat de fulmine nomen."
MARB., *Carm. de Gemmis*, § 30.

[3] Salm., Plin. Exercit., p. 179, a B.

[4] "Illis quippe locis, quos constat fulmine tactos,
Iste lapis tantum reperiri posse putatur."
MARB., *Carm. de Gemmis.*

derbolt (κεραυνὸς), this same virtue is ascribed.[1]

The iris seems to have been a very limpid, prismatic crystal of quartz, which received its appellation from its casting upon the walls of the chamber the colors of the rainbow when exposed to the sun's rays. Brongniart supposed the iridescent varieties of rock-crystal to be the iris; but Pliny speaks of the colors, not as seen in the gem itself, but as reflected from the walls of the chamber.

The zeros,[2] which resembled in appearance the iris, but did not in like manner exhibit the prismatic colors, may have owed that difference merely to the imperfection noticed by Pliny, a speckled vein that ran across the crystal. Corsi supposes it to have been smoky quartz, smoke topaz.

We come now[3] to Pliny's alphabetical and other concluding lists of minerals, as to the greater part of which we should in vain seek to determine what they were. In very many cases the Greek name, with Pliny's Latin in-

---

1 "Qui caste gerit hunc a fulmine non ferietur;
Nec domus aut villæ, quibus affuerit lapis ille."
MARB., *Carm. de Gemmis*, § 30.

2 Plin., H. N., xxxvii., 53.

3 Ibid.

terpretation of it, is the only trace left to guide our investigation; and we are as far from knowing what the minerals so designated were, as Theophrastus, if recalled to life, received in a modern cabinet, and presented with its catalogue, would be from recognizing idocrase, harmotome, and epidote merely from their names, although he might perceive that these names indicated certain characters or properties of the substances to which they are applied.

As to some of the minerals alphabetically arranged there is no doubt that they were organic fossils. Such was Hammonis cornu, "reckoned among the most sacred gems of Egypt; of a golden color, and in shape resembling a ram's horn."[1] And such appear to have been balanites, belus, and tecolithus;[2] bucardia; encardia, of which three varieties are specified; enorchis; euneos, resembling the kernel of an olive, striated like a shell, and not very white; idæi dactyli, of an iron color, and resembling the human thumb; meconites; nympharena,

[1] Plin., H. N., xxxvii., 60.

[2] These three are one and the same mineral, the lapis Iudaicus, called tecolithus from its fancied lithontriptic virtues. Dioscorides describes it (v., 155) as of a very regular shape, resembling an acorn, white, and with parallel lines traced upon its surface, as if in a turning-lathe.

like the teeth of the hippopotamus; ostracias, or ostracites,[1] so called from its resemblance to oysters (ostrea); phœnicitis; phycitis; and syringitis.

The fossil ivory that Theophrastus speaks of,[2] and the bony stones, and bones produced in the earth, which Pliny cites him as mentioning, were, no doubt, organic remains; as also the palmati, which Pliny, in the same passage, says were found near Munda, and that as often as you broke the rock.[3] Other substances mentioned in these lists of Pliny are not to be classed with minerals; as, for example, bronte, cinædia, corallis,[4] chloritis, dracontites, gorgo-

---

[1] Dioscorides, too, speaks of a stone, ostracites, resembling a shell (λίθος ὀστρακίτης, ὅμοιος ὀστράκῳ, v., 165), which was flat, divisible into laminæ, and used by women as a substitute for pumice in removing hairs. That may have been the same with this supposed fossil. He elsewhere (v., 84), however, calls a variety of furnace calamine ὀστρακῖτις, and, since Pliny in this alphabetical list speaks of a cadmitis, between which and ostracitis he makes but a slight distinction, it is possible they may be all of them one and the same thing, called by different names.

[2] De Lapid., c. 65.

[3] For the views of ancient authors respecting organic fossils, see Park., Org. Rem., vol. i., 15.

[4] Corallis, described as resembling vermilion, and brought from India and Syene, was probably red coral, as was also gorgonia, so called, Pliny says (xxxvii., 59), because changed

nia, hyænia, sauritis. Some minerals mentioned in these lists had been previously spoken of at greater length; as alabastrites, asbestos, capnitis, callaïs, coralloachates, obsidianus, chrysolithus, chrysoprasus, chalcitis, and hæmatites. Other names, again, are synonymous with those of minerals before examined; and not a few of these gems we may suspect to be, like the strange virtues and properties ascribed to them, wholly imaginary.

Here and there in the lists we recognize a mineral which still retains its ancient name, and is sufficiently well described to leave no doubt concerning it; as the heliotropium, a gem which occurred in Æthiopia, Africa, and Cyprus, of a leek-green color, with blood-red veins. Its name, according to Pliny, is derived from the manner in which it reflected the solar

---

to the hardness of stone; for the ancients supposed coral to grow as a vegetable underneath the waves, and to harden into stone when removed from its native element. Wherefore Ovid says,

> "Quo primum contigit auras
> Tempore durescit, mollis fuit herba sub undis."

In the Orphic poem (v. 511, *seqq.*) this transformation is described at length. See also Pliny, H. N., xxxii., 11. The Carmen de Gemmis of Marbodæus begins its account of coral thus (§ 22):

> "Coralius lapis est, dum vivit in æquore, vimen."

rays when immersed in water, and from its being used as a mirror in which to observe a solar eclipse.

But let us proceed to notice now, in Pliny's order, such other minerals in these lists as can be determined, or respecting which any probable conjecture may be formed.

The *achates* (agate), which stands first in order,[1] is spoken of by Theophrastus as a beautiful and rare stone, from the River Achates in Sicily, which sold at a high price;[2] but Pliny tells us that in his time it was, though once highly valued, no longer in esteem; it being then found in many places, of large size and very diversified appearance. He specifies eight varieties, which had appropriate names, derived from their respective colors, or from some other distinctive character.

That in which jasper was associated with agate, the jasper-agate of modern mineralogists,[3] was called iaspachates. Cerachates was of a wax color; a variety little valued, because of its abundance. Sardachates was composed in part of the sard or carnelian. Hæmachates, sprinkled with spots of jasper, or blood-red

[1] Plin., H. N., xxxvii., 54.
[2] De Lapid., c. 58.
[3] Cleav. Min., p. 267.

chalcedony, was the variety now called dotted agate. Leucachates was, as its name imports, of a whitish color. Dendrachates was our dendritic agate; a variety beautifully described in the Orphic poem under the name of ἀχάτης δενδρήεις.[1] Coralloachates was so called from some resemblance that it bore to coral. Marbodæus speaks of it as a gem containing saffron-colored veins;[2] and Pliny describes it as sprinkled, like the sapphire, with spots of gold. May they not, in this case, have confounded with agate yellow fluor-spar, containing, as it sometimes does, disseminated particles of iron pyrites? For that other minerals besides agate are comprehended in this list is very certain, since the next mentioned variety, autachates,[3] is described as diffusing, when burned, a fragrance resembling that of myrrh. It is not worth our while to consider, with Pliny, the medicinal virtues of this stone; but it deserves to be noted that he says physicians made small mortars of it.

Androdamas is described as possessing a silvery lustre, and a cubical form like that of dice. The same name has been previously given to a

[1] Orph., Lith., v., 230. [2] Marb., De Gemmis, § 2.

[3] Or, as Saumaise would read, stactachates.

variety of what Pliny calls hæmatite, which we found reason to regard as magnetic iron ore.[1]

That androdamas was described as of a black color, while this is said to have a silvery lustre (argenti nitorem), which seems to forbid our regarding them as the same mineral. We may, nevertheless, suspect some relationship between the two, since the one now under consideration is said by Pliny to resemble adamas; that is, probably, the magnetic oxyde of iron so called.[2] All circumstances being considered, we might venture to pronounce this androdamas magnetic pyrites, which, like the common sulphuret of iron, is sometimes found in cubes.[3] The lustre, indeed, is here said to be that of silver; but this false character, supposing it magnetic pyrites, may easily have been transferred to it from the arsenical pyrites with which it is sometimes found associated. And this latter is, perhaps, the mineral designated by the name which follows next in Pliny, *argyrodamas;* as to which it seems to have been undecided whether it was identical with or different from androdamas. There appears to be as little about this mineral as about the ὑαλοειδής of

[1] See before, p. 177. [2] See before, p. 193.

[3] Cleav. Min., p. 591.

Theophrastus to justify our regarding it as moon-stone.[1]

Antipathes was a name common to several plants and minerals thought to possess the virtue ascribed by the Magi to the substance here so called by Pliny, and which is supposed to be black coral.

Arabica has been already mentioned as Arabus *lapis*, while here it is Arabica *gemma*. It is spoken of by Dioscorides also,[2] and by Galen, and was probably a fine white marble.

Aromatites, said by some of Pliny's commentators to have been amber, seems to have resembled in its properties the fossil copal discovered at Highgate Hill, near London.

Augites was thought by many, Pliny says, to be different from callaïs. The inference to be drawn from this remark is, that generally it was regarded as the same with that mineral, which was probably turquoise.[3]

Amphitane, otherwise called chrysocolla, occurred in the form of a cube, and was said to possess magnetic properties; whence we might suppose it to have been magnetic pyrites in a cubic form. It may have been called chryso-

---

[1] See before, p. 227.

[2] Lib. v., c. 149.

[3] See before, p. 214.

colla from its supposed property of attracting gold as well as iron;[1] for it seems to have been a wholly different thing from the chrysocolla of Theophrastus and Pliny spoken of before.[2]

Aphrodisiace is supposed by Dalecamp to have been some variety of agate; but Pliny mentions a variety of amethyst called *Veneris gemma;* and the only description of the gem now in question, "ex candido rufa est," as well as its name, aphrodisiace, agrees with the belief that, like the gem of Venus, it was amethyst.

Ægyptilla seems to have been a name common to several varieties of agate. It was, perhaps, the ancient denomination of what is still called *Egyptian* pebble; a striped jasper; the quartz agatheon of Haüy.

Baptes[3] is thought, from its description and its name, to have been amber, dyed or stained of some other than its natural color.[4]

✓ *Beli oculus* appears to have been the mineral now called cat's eye.[5]

Bostrychites may have been amianthus.

---

[1] "Affirmaturque natura ejus, quæ magnetis: nisi quod trahere quoque aurum traditur."—PLIN., *H. N.*, xxxvii., 54.

[2] Page 81.

[3] Plin., H. N., xxxvii., 55.

[4] See before, p. 144.

[5] Boethius de Boot, Hardouin, Millin, and Corsi all agree in so considering it.

Bucardia is thought by Dalecamp and Hardouin to have been turquoise.

Catochitis, the same commentators think, was either amber, or some variety of bitumen.

Cepitis was, perhaps, an agate.

Chalcophonos, of a black color, and which, when struck, rang with the sound of brass, was perhaps clink-stone.

Choaspitis is, by Dalecamp, upon apparently slight grounds, supposed to have been chrysoberyl.

Chrysopis is, by the same critic, thought to have been hyacinth; and *eupetalos*, a variety of opal.

Lepidotis, which resembled the scales of fish, we might venture to call lepidolite.

Morochites, described as leek-colored, but its streak white—for so the words "lacte sudat" seem, by commentators generally, to be understood — was otherwise called moroxus; and hence has Karsten derived his name for asparagus-stone, which he calls moroxite. This morochites or moroxus we have already, when speaking of saline substances, found occasion to consider.[1]

---

[1] See before, p. 138.

Samothracia, which was black, light, and resembling wood, may have been lignite; and

Selenites, bating the fabulous part of its description, we may regard as the same mineral still called selenite.

Pliny concludes his treatise upon gems with some general remarks, and directions for distinguishing the true from false. The difficulty of doing which, he says, was great, because the false stones were sometimes made from true ones of a different kind. The sardonyx, for example, by cementing three gems of as many different colors, black, white, and red.[1] We find Theophrastus speaking of counterfeit emeralds; and Pliny, of beryl, opal, ruby, amethyst, and other stones, so well imitated that it was not easy to distinguish true from false.

The skill of the ancients in their manufacture of glass was such that they not only made

---

[1] H. N., xxxvii., 75. A similar fraud is sometimes practiced now by cutting, in like manner, colored glass and rock-crystal; giving to each one plain surface, and joining them by that with a colorless cement; then setting them in such manner that the colored glass and the joint by which it is united to the limpid crystal shall be inclosed within the metal. By means of these doublets, as they are called, may stones of any color be so well counterfeited that the fraud can scarcely be detected. (See Min. des Gens du Monde, p. 135.)

it of a crystalline purity;[1] shaped it by blowing; ground it in the lathe; and carved it like silver;[2] but the collection which Mr. Dodwell first formed and brought into notice at Rome, proves that they could successfully, and did commonly imitate, by means of it, almost every known marble and every sort of precious stone.[3] Beckmann speaks[4] of two ancient gems in the Museum Victorium at Rome, a chrysolite and an emerald; factitious, both of them, but perfectly well executed; perfectly transparent, and colored throughout, and free externally and internally from the smallest blemish.

The Portland (once called the Barberini) vase, by some regarded as a sardonyx, and by others, for a long time, supposed to be porcelain, is now ascertained to be glass. It is semi-transparent, of a deep blue color, with opaque white ornaments cut by the lapidary in bas-relief.

---

1 Plin., H. N., xxxvi., 67.

2 Beck., Hist. of Inv., iii., 208.

3 Gell's Pomp., vol. i., p. 98. Winckelmann asserts that the ancients had carried the art of working glass to a perfection which we are yet far from having equaled, and he appeals to works that still remain for the proof of his assertion. (See Stor. delle A. di Disegno, vol. i., p. 26, 28, 29.)

4 Hist. of Inv., i., 199.

The period at which glass was first invented is unknown, but the process of glass-blowing is represented in paintings at Beni-Hassan, executed 1700 years before the Christian era;[1] and a glass bead has been found in Egypt, bearing the name of a Pharaoh who lived 200 years later. Drinking-vessels of glass, of Egyptian manufacture, are repeatedly mentioned by Martial.[2]

Those who wish to inquire into the subject of malleable glass, to which Pliny is not the only ancient who alludes, will find the authorities collected under the proper head in Lardner's Cabinet Encyclopædia.

We had before occasion to remark that, even at the early period of their exodus, the children of Israel had among them engravers of precious stones. They, of course, derived their art from Egypt; in which country, and in India, we, at a later period, find that such stones were not only engraved, but imitated with consummate skill. From one or other of those countries Greece learned both these arts, and they were practiced in Pliny's time by his own countrymen also, though, probably, with less skill than

[1] Wilk., Anc. Egyp., iii., 89.

[2] Epig., lib. xiv., 94, 114, etc.

by the Greeks. Here again, therefore, do we find occasion to observe the difference between the scientific knowledge of these ancients and their ingenious art. They were fully persuaded that rock-crystal was permanent ice, and called it, accordingly, κρύσταλλος;[1] but, knowing thus little of its nature, they could, nevertheless, not only form beautiful vases, drinking-cups, and other vessels of it, many of which still remain to testify their skill, but impart to it various colors,[2] and produce from it such close imitations of the emerald and other gems that, Pliny says, there was no more gainful fraud practiced; and he declines to point out the authors from whom it might be learned, but prefers rather to suggest some criteria for distinguishing the false gems from the true.

One distinctive character, he remarks, is the superior weight of the true gem. Others are the uniformity of its structure, its freedom from vesicles within and from roughness on the surface, its lustre, and its hardness. He observes that fragments of obsidian will not scratch true

[1] See before, p. 187.

[2] Seneca (Ep. xix.) ascribes to Democritus the discovery "quemadmodum decoctus calculus in smaragdum converteretur;" which, Beckmann thinks, was by coloring rock-crystal. (Hist. of Inv., i., 198.)

gems; but that with the diamond (adamante) all of them may be engraved.

Of the multitude of stones we have seen classed as gems, but a small portion was made use of by the engraver. These were principally carnelians, and such others of the silicious kind as possessed the properties required for the nicest execution, without being too valuable in themselves to allow of having their weight and lustre impaired by the engraver's tool.[1]

To conclude now, it may be conceded by an admirer of the ancients that as naturalists they are not presented to us in a very favorable light; that their acquaintance with natural science was empirical—that of an artisan rather than of the philosopher. But they could not have practiced with success so many ingenious arts, unless they had known the properties of a vast number of bodies; and their ignorance in other points, where knowledge is thought indispensable to modern artists, only renders it the greater wonder that they were able to ac-

[1] On the art of polishing and engraving precious stones, one of the most curious and ingenious of all antiquity, see a learned chapter in Goguet's Origine des Loix, etc., tom. ii., p. 111. See also Winckelmann, Stor. delle Arti del Disegno, vol. i., p. 25; vol. ii., p. 20.

complish what they did. Pliny breaks into exclamations at the prodigious works achieved by man in his search for the precious metals in the bowels of the earth, comparing them to the labors of the giants. And well he may, since he has to speak of penetrating mountains, with the means which then could be employed, to a distance of near eight thousand feet, through flint,[1] and a substance which he describes as harder still than that to pierce. How slowly, if deprived of the aid of gunpowder, would our miners conduct such operations. But in this instance, as in many others, art and ingenuity supplied the place of tools; and one of the greatest philosophers of our own age[2] has acknowledged that the ancients were, in all pursuits which require only the native powers of the intellect or the refinements of taste, pre-eminent; that their literature and their works of art offer models which have never been excelled; that they possessed, as if instinctively, the perception of every thing beautiful, grand, and decorous; and that as natural philosophers they failed, not from want of genius or application, but merely because they followed a mistaken path.

---

[1] H. N., xxxiii., 21. Silicem is his word; but here, as often elsewhere, it may mean limestone.

[2] Sir H. Davy, Philos. of Chem., p. 4, Introd.

INDEX.

THE END.

Abbott's Juvenile Series.

The Little Learner.

A Series for very young children, in five small quarto volumes, beautifully illustrated, and designed to assist in the earliest development of the mind of a child, while under its mother's special care, during the first five or six years of life, as follows:

Learning to Talk;

Or, Entertaining and Instructive Lessons in the Use of Language. By JACOB ABBOTT. Illustrated with 170 Engravings. Small 4to, Muslin, 50 cents.

Learning to Think.

Consisting of Easy and Entertaining Lessons, Designed to Assist in the first unfolding of the Reflective and Reasoning Powers of Children. By JACOB ABBOTT. Illustrated with 120 Engravings. Small 4to, Muslin, 50 cents.

Learning to Read.

Consisting of Easy and Entertaining Lessons, Designed to Assist Young Children in Studying the Forms of the Letters, and in beginning to Read. By JACOB ABBOTT. Illustrated with 160 Engravings. Small 4to, Muslin, 50 cents.

Learning about Common Things;

Or, Familiar Instructions for Children in respect to the Objects around them that attract their Attention and awaken their Curiosity in the Earliest Years of Life. By JACOB ABBOTT. Illustrated with 120 Engravings. Small 4to, Muslin, 50 cents.

Learning about Right and Wrong;

Or, Entertaining and Instructive Lessons for Young Children, in respect to their Duty. By JACOB ABBOTT. Illustrated with 90 Engravings. Small 4to, Muslin, 50 cents.

Price of the Set, including case, $2 50.

Harper's Story Books.

A Series of Narratives, Biographies, and Tales, for the Instruction and Entertainment of the Young.

In Twelve quarterly volumes of 480 pages each, bound in blue; or Thirty-six monthly volumes of 160 pages each, bound in red. The whole Series illustrated with over One Thousand beautiful Engravings.

The volumes are of small quarto size, and are beautifully printed and bound. The Series is now complete.

Price of the set in quarterly volumes, including case . . .	$12 00
" " monthly " " " . . .	14 40
Price of each quarterly volume, containing three stories each	1 00
Price of each monthly volume, one story	40

Marco Paul's Voyages and Travels in Pursuit of Knowledge.

In Six volumes 16mo. These volumes present, in connection with a narrative of juvenile adventures, a great variety of useful information in respect to the geography, scenery, and customs of the particular places and sections of country visited, and are richly illustrated with engravings.

The subjects of the volumes are,

1. New York.
2. The Erie Canal.
3. The Forests of Maine.
4. Vermont.
5. Boston.
6. The Springfield Armory.

Price of the set, including case	$3 00
Price of each volume, separately	50

A Summer in Scotland.

A narrative of observations and adventures made by the author during a summer spent among the glens and Highlands in Scotland. Illustrated with Engravings.

Price $1 00

The Franconia Stories.

In Ten volumes 16mo. Each volume is a distinct and independent work, having no necessary connection of incidents with those that precede or follow it, while yet the characters, and the scenes in which the stories are laid, are the same in all. They present pleasing pictures of happy domestic life, and are intended not only to amuse and entertain the children who shall peruse them, but to furnish them with attractive lessons of moral and intellectual instruction, and to train their hearts to habits of ready and cheerful subordination to duty and law.

The following are the names of the several volumes:

1. Malleville.
2. Mary Bell.
3. Ellen Linn.
4. Wallace.
5. Beechnut.
6. Stuyvesant,
7. Agnes.
8. Mary Erskine.
9. Rodolphus.
10. Caroline.

The volumes are illustrated with numerous beautiful engravings.

Price of the set complete, including case $5 00
Price of the volumes separately 50

Young Christian Series.

Complete in Four 12mo volumes, richly illustrated with engravings, and beautifully bound.

1. The Young Christian.
2. The Corner Stone.
3. The Way to do Good.
4. Hoaryhead & M'Donner.

It is superfluous to speak of the rare merits of Mr. Abbott's writings on the subject of practical religion. Their extensive circulation, not only in our own country, but in England, Scotland, Ireland, France, Germany, Holland, India, and at various missionary stations throughout the globe, evinces the excellence of their plan, and the felicity with which it has been executed. In unfolding the different topics which he takes in hand, Mr. Abbott reasons clearly, concisely, and to the point; but the severity of the argument is always relieved by a singular variety and beauty of illustration. It is this admirable combination of discussion with incident that invests his writings with an almost equal charm for readers of every diversity of age and culture.

Price of the set complete, bound in Muslin $4 00
Price of the set complete, bound in Half Calf 7 40
Each volume separately, Muslin 1 00
Each volume separately, Half Calf 1 85

Harper's New Catalogue.

A new Descriptive Catalogue of Harper & Brothers' Publications is now ready for distribution, and may be obtained gratuitously on application to the Publishers personally, or by letter enclosing six cents in postage stamps.

The attention of gentlemen, in town or country, designing to form Libraries or enrich their literary collections, is respectfully invited to this Catalogue, which will be found to comprise a large proportion of the standard and most esteemed works in English Literature—comprehending more than two thousand volumes—which are offered in most instances at less than one half the cost of similar productions in England.

To Librarians and others connected with Colleges, Schools, etc., who may not have access to a reliable guide in forming the true estimate of literary productions, it is believed the present Catalogue will prove especially valuable as a manual of reference.

To prevent disappointment, it is suggested that, whenever books can not be obtained through any bookseller or local agent, applications with remittance should be addressed direct to the Publishers, which will be promptly attended to.

Franklin Square, New York.

www.ingramcontent.com/pod-product-compliance
Lightning Source LLC
LaVergne TN
LVHW010249110826
845151LV00004B/1424